Polyurethanes as Specialty Chemicals

Principles and Applications

Polyurethanes as Specialty Chemicals

Principles and Applications

T. Thomson

CRC PRESS

Boca Raton   London   New York   Washington, D.C.

Library of Congress Cataloging-in-Publication Data

Thomson, T. (Tim)
    Polyurethanes as specialty chemicals : principles and applications / T. Thomson.
        p.   cm.
    Includes bibliographical references and index.
    ISBN 0-8493-1857-2 (alk. paper)
    1. Polyurethanes—Environmental aspects. 2. Polyurethanes—Biotechnology. I. Title.

TP1180.P8T55 2004
668.4′239—dc22   2004049710

Visit the CRC Press Web site at www.crcpress.com

© 2005 by CRC Press LLC

No claim to original U.S. Government works
International Standard Book Number 0-8493-1857-2
Library of Congress Card Number 2004049710
Printed in the United States of America 1 2 3 4 5 6 7 8 9 0
Printed on acid-free paper

It is all well and good to copy what you see, but it is much better to portray what you can't see. The transformation is assisted by both memory and imagination. You limit yourself by reproducing only what has struck you, that is to say what is necessary. In this way, memory and imagination are freed from the tyranny exerted by nature.

Comment about impressionists attributed to Edgar Degas

Preface

It is traditional to begin books about polyurethanes by defining the class of polymers that has come to be known as polyurethanes. Unlike olefin-based polymers (polyethylene, polypropylene, etc.), the uniqueness of polyurethane is that it results not from a specific monomer (ethylene, propylene, etc.), but rather from a type of reaction, specifically the formation of a specific chemical bond. Inevitably, the discussion in traditional books then progresses to the component parts, the production processes, and ultimately the uses. This is, of course, a logical progression inasmuch as most tests about polyurethanes are written for and by current or aspiring PUR (the accepted abbreviation for conventional polyurethanes) chemists. Unlike discussions about polyolefins where the monomer, for the most part, defines the properties of the final product, a discussion of PURs must begin with the wide variety of constituent parts and their effects on the resultant polymers.

Thus, while ethylene defines the properties of polyethylene and vinyl chloride defines polyvinyl chloride, thousands of isocyanates and polyols define the polyurethane category. In olefin chemistry, differentiation is established by processing method. With polyurethanes, any discussion must cover both the process and the constituent parts. The flexibility thus conveyed permits their use in devices as diverse as skateboard wheels, dressings for treatment of chronic wounds, and furniture cushions. All of these items can be manufactured after minor changes are made in the chemistry. To cite another example, an ingredient change from polypropylene glycol to polyethylene glycol can restructure a business from one focused on furniture cushions to one focused on advanced medical devices.

This book will approach the subject of polyurethanes from an alternate point of view. While PUR chemists will find some new information, the target audiences for this book are the scientists and engineers who are in search of new material in the course of their research. These scientists are not from typical PUR disciplines. Some are environmental engineers looking for solvent extraction systems to remove pollutants from ground water. Some are engineers at municipal waste treatment facilities who must develop systems to remove H_2S from effluent air. Others are biochemists searching for a three-dimensional scaffold on which to grow cells.

The traditional markets for PUR are structural in nature. Furniture cushions and foam in general are the dominant forms of PUR. Automobile bumpers, shoe soles and inserts, insulation, and paints are also products of the chemistry and depend on physical properties of resilience and toughness. It is logical to begin this book with the definition of the chemistry and progress through the technology in the traditional fashion. It is paradoxical, however, that a chemistry that allows so many degrees of freedom is used so narrowly. Writing a book from the basis of the chemistry is, therefore, straightforward. The target (a polymer with a specific range of physical properties) is well defined. While a wide range of components can produce such polymers, the list of useful ones (considering availability and cost) is quite short.

Our approach to the chemistry of the polyurethanes has no such limitations, and we use it to some advantage. While we take advantage of the physical properties of PURs, our focus is on what happens to a fluid (gas or liquid) when it passes through or otherwise comes in contact with a polyurethane chemistry. It has been part of the polyurethane tradition to consider the material inert. By removing the traditional restraints of conventional raw materials and a limited range of end uses, we allow the chemistry to affect the fluid or components of the fluid.

However, we will not ignore physical properties. A section of the book will focus on structure–property relationships. PURs form devices that have chemical and physical features. The great value of polyurethanes as we will show in this book is the freedom to take advantage of their chemical and physical features and efficacies. While much of the book focuses on foams, we will also discuss coatings, membranes, elastomers, and their application to the problems addressed.

I must thank those who have molded our education in polyurethanes. Since the last book, my focus has moved from hydrophilic polyurethanes to more broad-based applications of this chemistry. While I still do not consider myself an expert in the field of PUR chemistry, I have tried to apply it to a broad range of practical uses and approach the subject from the perspective of a PUR researcher rather than as a manufacturer.

I want to thank my colleagues and investors for allowing me to spend my life playing around with this interesting "stuff." In this new adventure, they have not only listened to predictions and projections, they have supported them with time, energy, and money. Without them, I would be a security guard with a gun.

Lastly, I thank my wife, Maguy, whose support and love make me want to do better.

Table of Contents

About the Author

T. (Tim) Thomson, MS, is the director of Main Street Technologies, a consulting practice. He is also the chief technical officer of Hydrophilix, Inc. of West Newbury, MA, a technology-based firm specializing in the development of advanced medical devices, environmental remediation technologies and consumer products. He was the chief technical officer of Biomerix Corp. during its formative stages. Biomerix develops polyurethane-based drug delivery systems.

He is known worldwide for his expertise in the development of a broad range of products based on hydrophilic polyurethane and has authored a book on the subject. He has published a number of papers on the use of polyurethanes in medical and other applications. He has conducted seminars in the U.S. and Europe on the medical applications of specialty polyurethanes. He has been an invited speaker to a number of conferences and seminars.

Mr. Thomson began his career at Dow Chemical and held positions in manufacturing, research and technical support. He had assignments in the U.S. and Europe. He holds five patents in synthetic chemistry and process control. He has 11 patents applied for based on his development work with Hydrophilix.

His current activities include the application of polyurethane composites to the development of three-dimensional scaffold for cell growth (bacteria, plant and mammalian).

1 Introduction

It is traditional and typical for books on polyurethane chemistry to begin with a definition of a polyurethane, proceed to a listing of the component parts, and finally discuss the processes and design aspects. Despite the demonstrated versatility of polyurethane chemistry, its current applications, except for notable exceptions, are quite boring. Therefore, while most texts catalog the uses for the chemistry, the purpose of this text is to describe that chemistry — a subject only a chemist could love. The applications are noteworthy only as primers for design possibilities that, without exception, focus on the physical: making such polymers tougher, harder, etc. Polymer molecules are considered (or hoped to be) relatively inert. One of the purposes of this book is to dispel that notion. We will focus our attention on the chemical nature of the molecule and show that it can be used by researchers in a variety of disciplines. As we will show, the combination of the physical properties and chemical activities of polyurethane produces a remarkable partnership.

Before we get to the chemistry, it is important to mention that most polyurethanes are useful because of their physical properties, and the breadth of applications is remarkable. They can be stiff enough to be used as structural members and soft enough for cosmetic applicator sponges. They can serve as the wheels of inline skates or cushions for furniture. In these applications and hundreds of others, the chemistry can be summed up as a combination of hard segments and soft segments with varying degrees of cross-linking. This combination is, indeed, the strength of the chemistry. Changes in a limited number of component parts allow a wide variety of products to be made. It is therefore useful to discuss the subject from the perspective of its component parts and the processes by which they are combined, and we will do that in the next chapter.

Our task is more difficult than simply dealing with the physical properties. Not only do we have to be aware of and work with the structural parts of a polyurethane, we must also be able to effect changes in the molecule to create an environment that will exert effects on fluids (gases and liquids) with which the molecule may come into contact. As we stated, the polyurethane community generally regards the polymer as virtually inert.

The only other considerations are weathering, color development, and perhaps long-term oxidation. These are considered unfortunate problems to be minimized by various formulation techniques. In an extreme case, we all recognize that polyurethanes can be fire hazards, and this too must be addressed by various formulation technologies. In a sense, the slight reactivity of polyurethanes is considered a problem. We hope to show that opportunities arise from using the natural reactivity of the polymer surface and by making the polymer reactive to the environment with which it comes into contact.

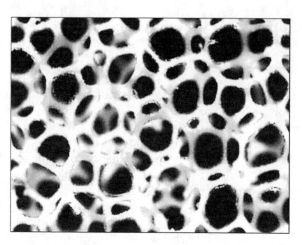

FIGURE 1.1 Reticulated polyurethane foam.

While this book covers the full range of polyurethane chemistries to one degree or another, our perspective has been on the chemical nature of the molecule. Unlike most polyurethane chemists, we have worked almost exclusively on hydrophilic polyurethanes. This specialty grade of polymer (which we will describe at length) is valued for its chemical properties (ability to absorb water, for instance) almost to the exclusion of its physical properties.

In recent years, we have become integrated into the much larger world of polyurethanes, but we have always begun our investigations with a focus on the surface chemistry. While our studies have been on the full range of polyurethane chemistries and the full range in which polyurethanes are produced, the chemical aspects in which we are most interested are foams (the bulk of polyurethane production), specifically open-celled foams, and more specifically products known in the industry as reticulated foams.

These foams are of special interest to us for several reasons. Chief among them is the high surface-to-volume ratio. The chemistry of the surface and the techniques we have developed to modify it best demonstrate the possibilities of the polymer to affect fluids passing through it. Other properties of interest are its strength, toughness, high void volume, and low pressure drop. Figure 1.1 is a micrograph of a typical reticulated foam. Many of the characteristics cited above are apparent in the picture. The realization that these properties are contained in a 2 lb/ft^3 package reinforces the qualitative impression.

An important theme of this book is impressing upon the reader the possibilities that are opened by adding aspects of chemical reactivity to the structure shown in Figure 1.1. In addition to describing how reticulated foams are produced and their physical parameters are varied, we will describe the ways we and others have used such structures. As noted area, the high surface area and low void volume make the reticulated foam a unique structure in material science.

Those of you who have worked with packed beds or beads are aware that the structures are produced by the interstitial spaces between the beads. In our work, we refer to reticulated foam as the "antibead."

A reticulated foam is the end result of a manufacturing technique as opposed to a chemistry. In the next chapter, we will introduce readers to the reactions and components that yield this class of polymer. It is important to note that most, if not all, of the foam formulations we will discuss can be converted into reticulated foams to take advantage of the properties of their unique structures.

Combining the aforementioned characteristics with a functionalized surface offers a product designer a unique platform for drug delivery, development of hybrid artificial organs, advanced plant growth media, and biofilters, flow-through solvent extraction, and a host of other applications. Put another way, while reticulated foam is essentially (but not entirely) inert, it exerts effects on the environment within its structure. The application of certain techniques can produce profound effects by changing the inert surface of the structure. Our work focuses on the fact that the reticulation process produces a unique scaffold that, when properly derivatized, can be nearly catalytic in its effect on fluids passing through it.

As stated, we will take a different route despite the tradition of beginning with a discussion of the molecules that constitute polyurethanes. We want to investigate the effects on the fluids that pass through or come into contact with polyurethane. In the simplest example, if air contaminated by polycyclic hydrocarbons passes through polyurethane foam, the concentration of hydrocarbons will change. In that sense, the foam is not truly inert. By the application of certain techniques, we will discuss how this effect can be controlled to provide an environmental remediation mechanism. We will discuss this effect in detail in this and subsequent chapters.

It is also a goal of this book to expand the audience to scientists and engineers who would not generally consult a book on polyurethanes to solve problems that arise in their professions. We want people to look at polyurethanes as possible solutions to their medical or environmental remediation assignments and go to polyurethane professionals for help.

In this first chapter, we seek to reinforce this perspective by including a series of case studies. We will propose problems in various areas of investigation and include specific examples of environmental remediation and advanced medical research issues addressed by polyurethanes in one form or another. While each example deals with a specific discipline, it is important to recognize that we have chosen all the examples in this chapter as surrogates with much broader applicabilities beyond the specific fields cited in the examples.

We will discuss the colonization of polyurethane by living cells. Two examples will be presented: one using bacterial cells and the other involving mammalian cells. The application of polyurethane technology is different for each situation but similar enough so that the reader will learn from both situations regardless of specific interest or responsibility.

In both cases, a nutrient solution (blood or polluted air or water) passes through and over cells and is changed by the action of the colony of cells. That action removes the toxin from a pollutant. The fact that the fluid passing through the polyurethane environment is a gas containing hydrogen sulfide or blood containing bilirubin is almost incidental (except to government officials who oversee the development of these technologies).

It is therefore important that readers look generally and specifically at the examples to determine applicability of the solutions to the problems they face in their work. The balance of the chapter is structured to propose a problem and then show how it was or could be addressed by the application of polyurethane chemistry. You will see that the solutions combine both the physical and the chemical aspects of the polymers. We will begin with an environmental problem of interest to both scientists and the general public.

AN ENVIRONMENTAL EXAMPLE

A natural and seemingly inevitable result of industrial development and human activity seems to be the release of organic and inorganic contaminants. We consume raw materials and release contaminants, often toxic, to the environment. Industrial development has led to the release of contaminants that range in toxicity from benign to acute to chronic. Agricultural progress, especially in the control of insects and weeds, has developed its own set of well-known pollutants. Most of these contaminants are handled naturally by the biosphere. Naturally occurring clays and rocks can remove many pollutants from water via ion exchange and adsorption processes. Bacteria, molds, and algae all have the ability to metabolize most pollutants. Septic tanks and municipal water waste treatment facilities depend on bacteria to degrade human waste.

When new pollutants are introduced into the environment, microorganisms in many cases evolve in order to use the contaminants as food sources. The concentrations of population in urban areas and large releases from industrial areas have in some cases outstripped the ability of the environment to handle the concentrations.

Certain synthetic organic pollutants have been designated as recalcitrant in the sense that the natural environment has not evolved a process to remove them. Halogenated hydrocarbons and certain pesticides are in this category. A recent report by the U.S. Geological Survey showed that population was a predictor of the probability of finding synthetic chemicals in potable ground water.[1] Figure 1.2 shows the probability of detecting volatile organic compounds (VOCs) in untreated groundwater across the U.S.

Treatments for this environmental problem range from physical methods and classic chemical processing techniques (distillation, extraction or sorption, for example) to biological treatments. Treatments in the latter category include *in situ* degradation using microorganisms and the direct application of enzymes. The use of a technology known as biofilters is of increasing interest. As we will show, both microbiological and chemical processing techniques benefit from the properties of polyurethanes.

In this first example, extraction of the contaminant from water is of particular interest for a number of reasons, not the least of which is that extraction requires no particular pretreatment of the contaminated fluid. Air can be injected into the soil around the aquifer and recovered in sorption towers for concentration and removal from the environment. Alternatively, the water can be pumped from the aquifer

FIGURE 1.2 Probability of finding VOCs in untreated groundwater in the U.S.

through extraction columns and reinjected into the groundwater system (assuming local regulations permit this).

In this context, *extraction* means any process by which a fluid (air or water) comes into contact with a material to which the pollutant has an affinity. The affinity can be a physical trapping modified by some form of surface energy or a solvent extraction process based on enthalpic principles. The result is that the fluid is pumped through the sorption medium and the pollutant is reduced or eliminated from the fluid. Despite limitations, the most common sorption medium is activated charcoal — a form of charcoal treated with oxygen to open millions of tiny pores between the carbon atoms. It is amorphous and is characterized by high adsorptivity for many gases and vapors.

The word *adsorb* is important here. When a material adsorbs something, it attaches to it by chemical attraction. The huge surface area of activated charcoal gives it countless bonding sites. When certain chemicals pass next to the carbon surface, they attach to the surface and are trapped.

Activated charcoal is good at trapping other carbon-based impurities (organic chemicals) and substances such as chlorine. Many other chemicals are not attracted to carbon — sodium, nitrates, etc. — and they pass right through a carbon-packed column. This means that an activated charcoal filter will remove certain impurities while ignoring others. It also means that an activated charcoal filter stops working when all its bonding sites are filled. At that point, the filter must be regenerated by reprocessing in steam.

For some applications, regeneration is not possible, and the material must be discarded. Additional problems include the fact that the charcoal sorbs based on molecular size; pollutants with molecular sizes greater than the pores of the charcoal are unaffected. Flow problems and attrition of the carbon particles are other difficulties. Activated charcoal columns are usually pressure vessels due to the large and dynamic pressure drops across the carbon bed.

Other extraction systems involve contacting a contaminated fluid (air or water) with a solvent for the pollutant. This requires a solvent that is environmentally acceptable (for example, biodegradable) or implementation of special precautions to ensure that the solvent is not released into the environment. Traditional solvents cannot be used for this purpose inasmuch as they are the contaminants that must be removed. A chlorinated solvent, even though it has ideal characteristics as an extractant, is a groundwater pollutant. Given the inevitable losses during the process, the result would be replacement of one pollutant by another.

While hundreds of materials probably could fulfill the broad requirements of a solvent for the extraction of pollutants, in this example we will start our investigation with work done at the University of Alabama on a process called *biphasic extraction*. Homopolymers and copolymers (referred to in this book as polyols) use components made from ethylene oxide (EO) and blends of ethylene oxide and propylene oxide (PO), respectively. Since they are soluble in water, they are not useful in solvent extraction schemes.

In order for a solvent extraction system to be of value, it must be able to separate the phase containing the pollutant from the water. While the polymers can be used to extract contaminants from air, their water solubility precludes separation from groundwater. In the biphasic technique, the separation of the polymer phase from the water is achieved by the well-known physical chemical effect known as *salting out*. Simply put, inorganic salts are added to the system. The addition has the effect of "dehydrating" the polyol, making it insoluble and permitting separation.

Part of their suitability is that these polymer systems are variable in molecular weight. At low molecular weights, they are water soluble, and as the molecular weight increases, the polypropylene glycol is water insoluble. The extra methyl group disrupts the ability of the polymer enough to prevent significant hydration. Thus, the result is that the polymer "solvent" can be adjusted to match the polarity (and therefore the solubility) of a pollutant by changing the ratio of EO to PO.

One of the most attractive features of this chemistry is that it is relatively benign, environmentally speaking. Therefore, while we may keep our minds open to other chemical systems, these polymer systems appear to be attractive solvents for remediation of contaminated water.

For the purpose of this argument, however, let us say that the biphasic system appears to be needlessly complicated. The reason for this might be the need for precise temperature control, not always possible in the field. Separation of the phases is possible but problematic on a large scale. Contamination by the use of inorganic salts to insolubilize the polymer precludes injecting groundwater back into the ground.

Other problems might include kinetics, contact area, polymer losses, and regeneration or disposal of the contaminated polyols. [Note: We are not suggesting that these problems are not addressed and mitigated by the fine researchers at Alabama. We are making a case for the development of an alternative.]

Thus, for logical or illogical reasons (perhaps even for commercial reasons), our hypothetical research team decided that, while it likes the use of the EO/PO polymer extraction technique, it wants to develop an alternative, but related, method. One strategy would be to insolubilize the polymer before it comes into contact with the polluted water. The strategy might be to add sufficient hydrophobic groups to prevent

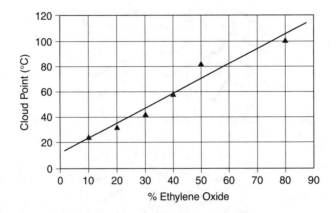

FIGURE 1.3 Effects of EO/PO ratio on cloud points of Pluronic surfactants.

the water from fully hydrating the backbone of the polymer. This is problematic in that as it would affect the ability of the polyol to extract.

This problem also exists with the biphasic system. In both strategies, a compromise would have to balance extractability and solubility. An example is to produce copolymers of EO and PO. At high concentrations of PO, the polymer becomes insoluble, but at the expense of decreasing the copolymer's ability to extract highly polar pollutants. The old rule that "like dissolves like" applies. The problem is mitigated, but not eliminated, by the construction of a block copolymer. When the concentrations of the PO and EO are adjusted just below the solubility level, small changes in temperature typical in field extraction studies can transform a system from soluble to insoluble.

A number of surfactant systems represent examples of the effect of the EO/PO ratio. Most notably is the Pluronic series of surfactants (Wyandotte Division of BASF Chemical, Wyandotte, MI). These surfactant systems are copolymers of the two oxides. Molecular weight is also an important consideration in their design.

One of the important quality control parameters is the *cloud point* — the temperature at which a solution of the polymer changes to a suspension or vice versa (see Figure 1.3). An examination of Pluronic product literature shows the effects of both EO/PO ratio and molecular weight. Since we recognize that these effects also impact solubility, we chose to look elsewhere for an answer; controlling all these factors in the field might be problematic.

At this point our team identified a well-known chemistry that shows great promise in combining the extractive properties of a water-soluble polyol in an insoluble polymer form. The polymer has the ability to be made into a number of physical conformations including films, membrane beads, and foams. Technology allowed us to graft this polymer onto a scaffold that provided physical strength, high surface area, high void volume, and certain valuable flow properties (e.g., low pressure drop).

All polymer chemists know this technique as *cross-linking*. It is the process of building intermolecular bridges. If two adjacent polymer molecules of equal size are connected by a cross-link, the molecular weight effectively doubles. From a

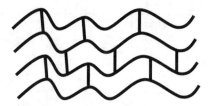

FIGURE 1.4 Random cross-linking of polymer.

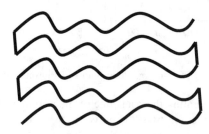

FIGURE 1.5 End-to-end cross-linking of polymer.

physical view, the effective molecular weight is increased to a point where water does not have the ability to hydrate the molecule fully and therefore cannot dissolve it. The process is actually gradual. As the molecular weight increases, the formerly soluble polymer makes a cloudy solution.

At higher molecular weights, the polymer begins to separate from the solution. This effect is similar in nature to changing the EO/PO ratio as described above — with one important difference. If these cross-links are placed randomly throughout the polymer backbone, the natural chemistry of the polymer is altered. Figure 1.4 depicts random cross-linking. If this principle is applied to the current problem, the ability of the polymer to extract is affected.

If, however, the cross-links are applied only to the ends of the polymer molecule, molecular weight increases still result in insolubility, but the character of the original backbone is maintained (at least partially). Figure 1.5 illustrates end-to-end cross-linking. Inasmuch as the purpose of this effort is to maintain polymer characteristics (ability to dissolve pollutants), it is logical that the goal should be to increase molecular weight by cross-linking at the ends of the molecule.

The nature of EO/PO polymers is to end in hydroxyl (also known as alcohol) groups. Thus, in the trade, the polymers are known as polyalcohols or polyols for short. End group cross-linking must be conducted at these alcohol end groups. While many chemistries are known to react with alcohol groups specifically, one stands out as particularly useful due to reaction rate, availability, cost, and ease of use.

The chemistry product is known as an isocyanate, and its reaction with a polyol is the basis for what we refer to as polyurethane. We will discuss the chemistry in detail in the next chapter, but for now it is sufficient to say that by reacting the polyol that best fits our needs from an extraction point of view with an isocyanate, we produce a water-insoluble system. Further, because of the way we constructed the polymer, it maintains much of the extraction character in which we are interested.

Another issue is worthy of note. One of the themes of this book is the juxtaposition of chemistry and geometry (physical characteristics). By the reaction of cross-linking the polyol, we are offered the opportunity (but not the obligation) to simultaneously produce a foam. This foam can then be reticulated to produce a unique combination of a solid solvent extractant in the form of a high surface area, flow-through medium.

Polyurethanes currently are not made to serve as solvent extraction systems. They are produced, as we have discussed, by design factors that focus on physical strength and form. Thus, our research team had to seek the help of polyurethane chemists to build the polymer to specifications that concentrate on its use as an extractant.

The current library of polyurethanes has some utility, and we will illustrate their uses with examples from our laboratory and from others. Currently, hydrophobic polyurethanes can be used to extract nonpolar pollutants, for example, from some pesticides. At the other end of the spectrum, hydrophilic polyurethanes can be used to extract sparingly soluble organic pollutants from groundwater. We will illustrate this with the extraction of methyl-*tert*-butylether.

To summarize this example, we have shown how a team of researchers might be led to polyurethane as an extraction solvent for aqueous-based pollutants. The polymer has attributes that can provide for additional benefits as well-including cost, surface area, and flow-through characteristics. This specific example deals with extraction from water, but many of the same arguments could have been applied to extractions from gases.

ANOTHER ENVIRONMENTAL APPLICATION

The biological treatment of contaminated water is prehistoric. One could say that the treatment is a natural process of recycling. Part of the system involves the accumulation of water in ponds and lakes followed by the growth of carbon-eating microorganisms. The latter is a process of natural selection. In modern times, this model is used to treat water contaminated by the concentration of populations and industrial development. While the mechanism is the same, modern systems are set up to handle increasingly large loads.

Microorganisms, yeasts, molds, bacteria, and algae are all parts of the natural process that reduces organic species back to their elemental units. The organisms consume organic components as fuel in the same sense that we consume cheeseburgers. The organisms then convert pollutants to energy and biomass. In part, the degraded pollutants are converted to CO_2 and water, and a portion becomes the biomass of the next generation of bacteria. The fortunate thing about bacterial and other lower species is that they are much less particular about what they eat. In fact, bacteria quickly evolve to develop the ability to consume prevalent organics in their environments.

As noted earlier, population density and/or industrial development can outstrip the ability of the natural environment to handle the occasional large amounts of

FIGURE 1.6 Municipal waste treatment facility at Saco, ME.

organics pumped into a system. The most visible consequence is the building of large municipal waste treatment facilities to accelerate the processes by which human organic waste is recycled back into the environment (Figure 1.6).

These processes are essentially accelerators to the natural process. At times, new molecules that have no corresponding organisms are developed. These molecules are referred to as *recalcitrants* and are of increasing interest to those who work with the environmental infrastructure.

The field of study that encompasses this technology is called bioremediation. From a practical view, the degradation of a wide variety of organic molecules is an accepted method. Microorganisms have developed to handle most common pollutants. Thus, municipal waste treatment plants operate without special needs for particular organisms.

The development of pesticides, for instance, led to the development of genetically engineered organisms and enzymes to handle particularly difficult or uncommon organics. Organophosphate pesticides (malathion, diazinon, parathion, etc.) are examples that will be discussed in detail in subsequent chapters. To develop this technology commercially, however, many of the microorganisms must be isolated because they are expensive or are not safe to release to effluent streams.

The technologies have, therefore, developed in the direction of attaching or encapsulating the organisms in a matrix. This technique is known as immobilization. Using this technique, a colony of organisms can proliferate on a substrate while an

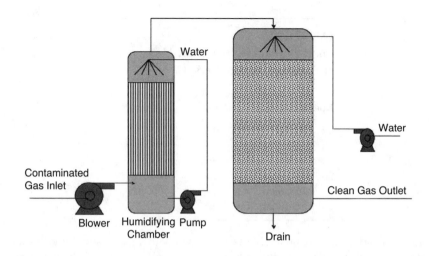

FIGURE 1.7 Typical trickling biofilter.

air or water stream containing a nutrient that is the target of the bioremediation can pass through it.

In the next example, our research team is asked to develop a system to minimize gaseous emissions from a plant. Parenthetically, these emissions are rarely illegal and for the most part represent an annoying but ubiquitous problem. However, the future holds every indication that emissions of this type will be made illegal. Additionally, other odorous emissions will have to be held within the boundaries of many facilities including food processing and chemical manufacturing plants. Therefore, while this example deals with a specific gas-phase pollutant, it represents a large number of other problems.

For reasons that will be explained in a subsequent chapter, the team found that the application of a technology known as *biofiltration* was considered the best remediation method. The contaminated gas is pumped through a column packed with a material that performs two functions: (1) it provides for efficient contact between the air passing through it and water, and (2) it serves as a scaffold for the development of a bacterial colony. Figure 1.7 shows a typical biofiltration arrangement. The team must develop a system that has a very low pressure drop so that the process can be constructed from plastic tanks and low-pressure pumps to save money. In a further effort to save money, the process should have a small footprint.

The bulk of the odor-causing gas from municipal waste treatment plants is hydrogen sulfide. Thiobacillus is a naturally occurring microorganism that consumes hydrogen sulfide and degrades it to SO_2 and water. The object of the study, however, is to find a suitable packing material that would serve as a support system for the growth of the organisms. Table 1.1 lists common packing materials.

These materials are commonly used in the chemical processing industry as packing for extraction and distillation columns. In a subsequent chapter, we will discuss the relevance of void volume, but it relates to the size of the equipment. A higher void volume is a more efficient use of usable space in a column. It is clear

TABLE 1.1
Common Packing Materials Used in Commercial Biofilters

Type of Material	Surface Area (M^2/M^3)	Void Volume (%)
0.5" Rock	420	50
0.5" Carbon	374	74
0.5" Berl	466	63
1.0" PVC saddle	249	69
1.0" PVC pall ring	217	93
1.0" Raschig ring	190	73

that a high surface area is preferable because the microorganisms will use this area to develop. A factor not typically addressed in the design of a biofilter is the relationship of the organism with the surface on which it is to reside. The materials listed in Table 1.1 are essentially inert (neither antagonistic nor beneficial) to organisms.

For the sake of argument, let us say that these column-packing materials are not sufficient due to size limitations. Additionally, the surface must be supportive (in the physical sense) and beneficial to the organism. It is desired to increase the strength of the attachment so that a high flow rate could be used through the column. The effect would be to increase the pressure drop through the column, and this would have to be accounted for in the packing material.

Because Thiobacillus "feed" on sulfur-containing compounds, they require other carbon-containing fuel to survive and multiply. Having a scaffold that can serve as a reservoir for nutrients could be an advantage, as opposed to the feed-and-starve cycle typically used. The team decided that the packing materials listed in Table 1.1 were not sufficient for the new biofilter design.

After an extensive review of possible new materials, the team found a material that had a surface-to-volume ratio closer to 1000 M^2/M^3 and a void volume up to 98%. A hydrophilic coating could be grafted to its surface to provide a reservoir capacity to release nutrients in a controlled manner. Lastly, the hydrophilic coating could be copolymerized with certain bioactive polymers and ligands that improve cell adhesion dramatically.

This new material is, of course, the reticulated polyurethane foam discussed in the earlier example. Later in this book, we will expand on this example to show how we and others used the chemistry and physical structure of polyurethane to remediate environmental pollution and also as a system to produce fine chemicals and proteins including enzymes for industrial and medical uses.

In these two examples, we described polyurethane as a physical device possessing such important features as a high surface-to-volume ratio and a high void volume. We also talked about it as a chemical system for solid solvent extraction and as a polymer system for enhancing the adhesion of cells. We will go into much more detail, but we have begun the process of considering polyurethane for uses beyond furniture cushions.

IMMOBILIZATION OF ENZYMES

Enzymes are naturally occurring proteins that have the ability to catalyze chemical reactions. All organisms use enzymes to break down food sources so the degraded forms can be absorbed into the cells as food. Enzymes break down carbohydrates to produce glucose. The cells use glucose as part of the energy system. Proteins are broken down by protease enzymes that are also used by the cells. In recent years, enzymes isolated from the microorganisms that produce them were used to address environmental concerns. An example that we will explore later is an enzyme produced by genetically engineered *Escherichia coli* and used to degrade organophosphate pesticides in agricultural runoff. It is harvested and used in remediation of contaminated surface water.

It is desirable under certain circumstances to use an enzyme in what is called an immobilized form. The enzyme is attached covalently or by entrapment in a polymer matrix. A contaminated fluid that comes into contact with the polymer is thus acted upon by the enzyme to produce a desired effect. We will discuss the advantages in subsequent chapters. While this technique lowers the efficiency of the enzyme, it extends its useful life by orders of magnitude, and the enzyme is not "thrown away with the bathwater."

The immobilization process, however, can be complicated. The immobilization of an enzyme on nylon serves as a point of comparison. The first step is to activate the surface of the nylon by treating it with hydrochloric acid at room temperature for 24 hours. The partially hydrolyzed nylon is then dried in ether and stored in a desiccator overnight. The nylon is then mixed with a coupling agent [1-ethyl-3-(3-dimethyaminopropyl)] and shaken for 1 hour. The enzyme is then added and shaken overnight at 4°C.

It is clear to the most casual observer that this technique is not suitable for large-scale production. The raw materials would be prohibitively expensive. In fairness, other immobilization techniques exist, and many are less formidable. None, however, has become dominant as a production technique.

The goal for our team of researchers is just that, however: to develop an immobilization technique that is economical, scalable to production-size equipment, and accomplishes its task with commonly available raw materials. We already discussed the reaction of isocyanate, a component in all polyurethanes, with alcohols. It is well known, however, that isocyanates also react vigorously with amines, carboxylic acids, and other moieties.

Since all proteins contain both of these reactive groups, if there were a possibility of producing a polyurethane, and particularly a reticulated polyurethane, with excess isocyanate groups, it would be possible to produce an enzymatically active surface on a high-surface-area, high-void-volume reticulated structure. This is possible and in fact is easier than the most common methods used currently to immobilize enzymes.

Consider the procedure for immobilizing an enzyme using polyurethane technology. A solution of the enzyme is produced in water. The solution is then emulsified with a hydrophilic polyurethane prepolymer. The emulsion is applied to the structural members of a reticulated foam by means of nip rollers. After curing for 15 minutes, the system is ready for use. Again, we have worked our way to a conclusion that a properly formulated polyurethane can be an effective solution through combining the geometry of reticulated foams and the chemistry of isocyanates.

A MEDICAL EXAMPLE

The mammalian liver is a construction of living cells that function (unlike in other organs) in a delicate choreography that simultaneously detoxifies, metabolizes, and synthesizes proteins. The liver handles the breakdown and synthesis of carbohydrates, lipids, amino acids, proteins, nucleic acids, and coenzymes (Figure 1.8).[2] In addition to the hepatocytes, other cells within the liver perform other vital functions. The system contributes to the disposition of particulates carried by the bloodstream and fights myriad microbiological agents responsible for a number of infectious diseases.[2]

The liver is interconnected with other organs (pancreas, spleen, intestine) along the portal venous circulatory system. Thus, it is clear that an *ad hoc* view of the liver function of removing toxins is insufficient in light of the other duties it performs.

Fulminant liver failure results from massive necrosis of liver tissue. Diminution of mental function results, and this often leads to coma. The body undergoes a buildup of toxic products, alteration of its acid balance, and a decrease in cerebral blood flow. Impaired blood coagulation and intestinal bleeding occur as well. Other malfunctions and diseases of the liver include viral infections and alcoholic hepatitis. In 1999, of the 14,707 individuals on a waiting list for transplants, 4,498 received transplants and 1,709 died while waiting.[3] As of February 2002, 18,434 people awaited liver transplants.

Partial and whole liver transplantation is considered the treatment of choice, but the need exceeds the supply. A device able at least temporarily to perform the

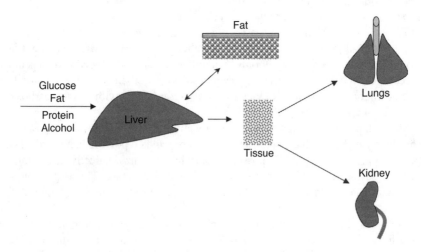

FIGURE 1.8 Human liver function. (Adapted from Greisheimer, E.M.[4])

functions of the liver would bridge the gap until suitable donors became available. It could potentially reduce the load on a compromised liver until regeneration restored full function and offered some hope to patients ineligible for liver transplants due to infectious disease or other complications. Most attempts to duplicate liver function, however, have failed. Dialysis techniques and transfusions have shown little long-term success.

Despite these failures, the need still exists for alternative liver therapies. Several new techniques including cell transplantation, tissue-engineered constructs, and extra- and paracorporeal devices seek to relieve some of the demands placed on a compromised liver. Liver assist devices allow the liver to regenerate its function by removing some of the demands. Bridge-to-transplant devices seek to maintain patients until suitable donors are available. Some therapies seek to remove toxins from the blood, and they have a place in the treatment scheme, but due to the complex and multifunctional nature of the organ, some type of cell-based therapy is considered a more complete solution.

Fortunately, advances in liver cell biology have provided valuable insights into the functioning of the organ. One such insight is the fact that hepatic cells function only when they are able to form spheroid structures. That means a flat plate of cells cannot function as an artificial liver. Liver cells are functional only when they can build three-dimensional structures (spheroids). The next generation of devices must bring together our current understanding of the biology of hepatic cells along with the cellular microstructures and the structures in which cells will grow. A structure known as a scaffold has been the focus of much research. Other strategies include hollow fibers and encapsulated spheres.

Our hypothetical team of medical professionals is charged with developing a liver assist device based on the development of a scaffold for the growth of hepatic spheroids. The device must have a high surface area and permit the migration of hepatic cells during the development of a large enough colony to support the patient. Hepatic cells are anchorage dependent — they must attach themselves to something in order to function. This characteristic must also be considered in determining the chemistry of the scaffold. Additionally, the product must be compatible with the cells and with blood passing through it. Devices currently under investigation do not include this feature. In those systems, the plasma is extracted from the blood and is treated by the artificial liver. The scaffold must have a significantly large void volume to accommodate the development of a hepatic colony yet still have sufficient excess volume to permit the infusion of blood without producing too much resistance to flow.

Another controversial aspect is the durability of the system. It has become common to find researchers who focus their attention on so-called biodegradable scaffolds. We have chosen as a design requirement of our hypothetical device to focus on so-called biodurability. This is not simply a construct for this example. It is a viable alternative that should be explored and will be discussed elsewhere.

Readers will understand that we are loading our case in favor of a reticulated polyurethane as the material of choice. Figure 1.8 qualitatively speaks to a structure that is conducive to the migration of a developing hepatic colony. The void volume and open structure discussed earlier contribute to a design aspect we call a "pseudo-vasculature."

Our proposal is not theoretical. Researchers have used reticulated hydrophobic polyurethanes as liver assist devices with some success.[5] We will discuss this research and future work in detail later. For now, it is useful to present an overview. Matsushita et al. inoculated a reticulated polyurethane with porcine hepatic cells. The device functioned as noted, but it was necessary to separate the plasma from the blood because conventional hydrophobic polyurethanes are not hemocompatible. In addition, the technique made no provision for cell attachment. Workers in our laboratory grafted a hydrophilic polyurethane to the structural members of a hydrophobic reticulated foam in an effort to make the composite hemocompatible. Additionally, this gave us the opportunity to add cell attachment proteins.

We have described the need for a device that could assist a compromised liver or even serve as a bridge until a transplant became available. We have compared the properties of an ideal scaffold for such a device with the structure of a reticulated foam and reported results of research into its use. Lastly, we have postulated improvements in current research that could lead to an efficacious solution.

SUMMARY

This chapter introduces readers to the versatility of polyurethane polymers without spending too much time on the chemistry. The next chapter will discuss a more classical view of the molecule and how it is developed. Our point, however, is to present a functional view of this system. We have examined its physical characteristics, focusing our attention on the uniqueness of reticulated foams. All the chemical points we have made apply to all polyurethane polymers, whether they are open-celled foams, closed-cell foams, or thermoplastic elastomers.

We have also cited polyurethanes as a chemical species with profound and sometimes subtle effects on the environment. The abilities to extract hydrocarbons and to serve as a surface for the colonization of cells were discussed as examples. In later chapters, we will discuss how these rather subtle features are amplified to produce easily recognizable properties.

We could have chosen examples involving drug delivery, agriculture, aquaculture, and production of cosmetic and personal care items with equal force and conviction. It is important for practitioners of those disciplines to continue reading this text and look for relevant applications.

As noted, most commercial polyurethanes are useful because of their physical properties. Except in the field of hydrophilic polyurethanes, little work has been done on the chemistry of polyurethanes. We hope this book will change that to a degree. Until then, however, basic research in this area will require the production of your own polymers.

2 Polyurethane Chemistry in Brief

The first chapter followed a nontraditional path in polyurethane chemistry. We first assumed the role of an environmental chemist seeking to develop a solid extraction solvent to remove pollutants from water and air. We chose to use polyalcohols because of the spectrum of polarities they possess. Using ethylene and propylene glycol, one can "design" an extractant system by varying the amount of each compound in a block or random polymer.

We were supported in our thesis by the work of Huddelston et al.[6] In order to achieve phase separation, they employed a "salting out" principle with good success. We employed a polyalcohol backbone and achieved phase separation by reacting the terminal alcohol groups with isocyanates. This was done after the addition of crosslinking chemicals yielded a solid polyol, which, as we will show in subsequent chapters, has the extractive properties we sought. Additional processing techniques allowed us to build an open-cell structure that permitted the flow of fluids and extracted the pollutants.

We then assumed the role of a team of medical device researchers who wanted to build a three-dimensional structure on which to propagate attachment-dependent cells. Several requirements were parts of the critical path of the project. First, the structure had to have high void volume and high surface area. It had to be biodegradable and produce nontoxic degradation products or it had to be biodurable. It also had to be biocompatible, preferably neutral, in order to grow cells. The polymer had to provide binding sites for the attachment of cells. Lastly, the material was to be hemocompatible so that it would not initiate inflammatory responses. Again, we resorted to the use of polyols, specifically, polyethylene glycol. The structural properties were achieved by reacting isocyanates to the polyols. The structural properties were achieved by an open-cell foam structure with 94% void volume, 300 M^2/M_3, with sufficient strength to withstand the environment in which it would have to reside.

These two examples show that a polyurethane — the reaction product of a polyol and an isocyanate — can serve in both geometric and chemical functions. This is essentially the theme of this book. Most polyurethanes are used for purposes other than the applications cited above. The true place of polyurethanes in the world today is based on the physical properties of the chemistry. What we hope to do is describe the polymer as a chemistry product with properties that are of use to scientists of various disciplines.

This chapter and the next will follow a more traditional pattern and describe this class of compound from a polymer chemist's point of view. While a number of more comprehensive texts on this subject are available, we will try to approach the subject not as pure polyurethane chemists, but as researchers experienced in the use of polyurethanes for their physical and chemical characteristics. We will try to provide the readers with a general view of current polyurethane practice. More importantly however, the purpose of this chapter is to teach the skills readers need to build polyurethanes of their own design. We will show that the commercially available polyurethanes are probably sufficient to fulfill the physical needs of most projects. However, in keeping with the theme of this book, it may be necessary to build a new class of polyurethanes with nonstandard polyols, for instance. The information in the chapter will give readers a starting point for manufacturing new polyurethanes. We will limit our discussion to commercial raw materials rather than attempting to cover the full range of possibilities. It is hoped that the information will assist researchers in identifying particular chemical aspects (biocompatibility, polarity, etc.) and building the aspects into a polyurethane of the necessary physical structure (foam, film, etc.) to solve a particular problem. This chapter describes the chemicals; the next discusses a series of structure–property relationships that will be useful.

PRIMARY BUILDING BLOCKS OF POLYURETHANE

ISOCYANATES

For those familiar with polymer chemistry, *polyurethane* may be a confusing term. Unlike polyethylene, the polymerization product of ethylene, a polyurethane is not the result of the polymerization of urethane. To add to the confusion, a urethane is a specific chemical bond that comprises a very small percentage of the bonds of a polyurethane. Since we are interested in chemical and physical effects, *polyether* or *polyester* is a more descriptive term for the most common bond in a polyurethane. Despite this complication, it is instructive to begin by talking about the urethane bond from which the *polyurethane* name is derived. The general structure or bond that forms the basis of this chemistry is the urethane linkage shown in Figure 2.1.

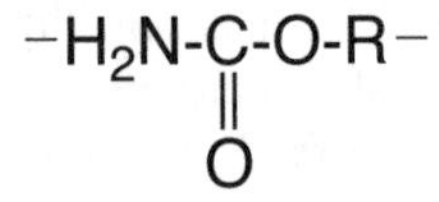

$$-H_2N-\underset{\underset{O}{\|}}{C}-O-R-$$

FIGURE 2.1 Basic urethane

The first urethanes involved the reaction of isocyanate with simple alcohols and amines. They were of sufficient economic value to foster the development of a number of isocyanates, including the aromatics that play dominant roles in modern polyurethanes. An isocyanate is made by reacting an amine with phosgene.

Isocyanates of the general structure shown in Figure 2.2 react vigorously with amines, alcohols, and carboxylic acids. Examples will be discussed later in this chapter. Table 2.1 presents the relative reaction rates of some of the molecules we will use to design our own polymers.

It was not until the 1930s that the work of Caruthers in the U.S.[7] and Bayer in Germany[8] led to the development of polymers based on diisocyanates and triisocyanates. The first polymers were based on diamines; the technology quickly shifted toward polyethers and polyesters. Isocyanate development after the 1930s has focused on aromatic isocyanates, more specifically on two molecules and close

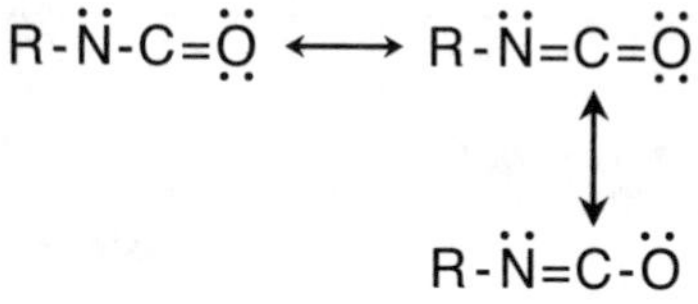

$$R\text{-}\ddot{N}\text{-}C\text{=}\ddot{O} \longleftrightarrow R\text{-}\ddot{N}\text{=}C\text{=}\ddot{O}$$

$$\updownarrow$$

$$R\text{-}\ddot{N}\text{=}C\text{-}\ddot{O}$$

FIGURE 2.2 Resonance structures of isocyanates.

TABLE 2.1
Relative Reaction Rates of Isocyanates

Active Hydrogen Compound	Relative Reaction Rate
Primary amine	100,000
Secondary amine	20,000–50,000
Water	100
Primary alcohol	100
Secondary alcohol	30
Carboxylic acid	40

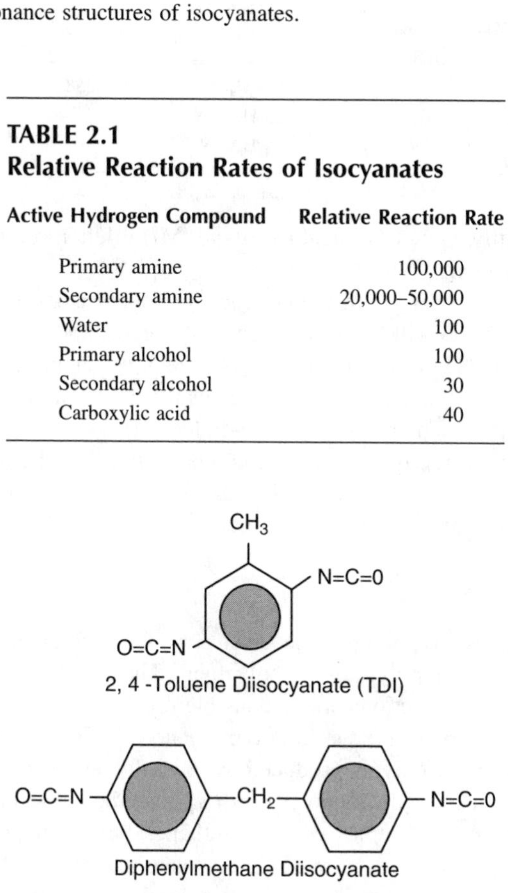

2, 4 -Toluene Diisocyanate (TDI)

Diphenylmethane Diisocyanate

FIGURE 2.3 Most common commercial diisocyanates.

variations. The molecules are toluene diisocyanate (TDI) and its isomers and meth-ylene-*bis*-diphenyl diisocyanate (MDI) in monomeric and polymeric forms (see Figure 2.3). As a general rule, TDI makes flexible polyurethanes and MDI produces stiffer polymers.

MDI offers a number of advantages. First, it is somewhat safe to use based on its much lower vapor pressure and is available in convenient forms. It is produced by the reaction of an amine and phosgene. The result is a mixture of multi-ring isocyanates. The purest form is the two-ring isomer shown in Figure 2.2. The isomer is recovered by distillation. What is left behind is the so-called polymeric MDI that

TABLE 2.2
Commercially Available Isocyanates

Designation	Formula	Molecular Weight	Melting Point
2,4-Toluene diisocyanate (TDI)	$C_9H_6O_2N_2$	174.2	21.8
Diphenylmethane-4,4'-diisocyanate (MDI)	$C_{15}H_{10}O_2N_2$	250.3	39.5
1,6- Hexane diisocyanate (HDI)	$C_8H_{12}O_2N_2$	168.2	–67
Hydrogenated MDI	$C_{15}H_{18}O_2N_2$	258.3	30
Isopherone diisocyanate (IPDI)	$C_{12}H_{18}O_2N_2$		–60
Naphthalene diisocyanate (NDI)	$C_{12}H_6O_2N_2$		127

is sold commercially. Dow Chemical (Midland, MI) offers more than a dozen pure and polymeric forms of MDI.[9]

It is important to be aware of the chemical effects of isocyanates. The polyurethanes you will develop will be combinations of polyols and isocyanates. The ratio of the two compounds will in part dictate both the physical and chemical properties of the product. As a general rule, the isocyanates are hard segments that impart rigidity to the polymer. The polyol is the so-called soft segment. The various molecular weights (more correctly equivalent weights available in the form of polymeric MDIs) provide certain advantages. Table 2.2 lists a few commercially available polyisocyanates and their physical properties.

POLYOLS

Since the early days of polyurethane discovery, the technology has focused on isocyanate reactions with polyesters or polyethers. The differences will be discussed in later sections. These reactions are responsible for the growth of the polyurethane industry. The polyesters of interest to polyurethane chemists terminate in hydroxyl groups and are therefore polyols produced by the polycondensation of dicarboxylic acids and polyols. An example is a polyol with a polycarbonate structure (Figure 2.3).

We will cite more examples of polyurethanes based on polyethers than on polyesters. The polyethers are more easily designed when the polarity of the backbone is important. For instance, one can use polyethers to construct polyurethanes that are hydrophilic or hydrophobic or react to water at all levels between these extremes. Polyethers permit the development of biocompatible and hemocompatible devices. Lastly, they are more hydrolytically stable and so are more appropriate for environmental studies.

$$HO-[-R-O-\overset{\overset{\textstyle O}{\|}}{C}-O-]_x-R-OH$$

FIGURE 2.4 Polycarbonate structure of polyester polyol.

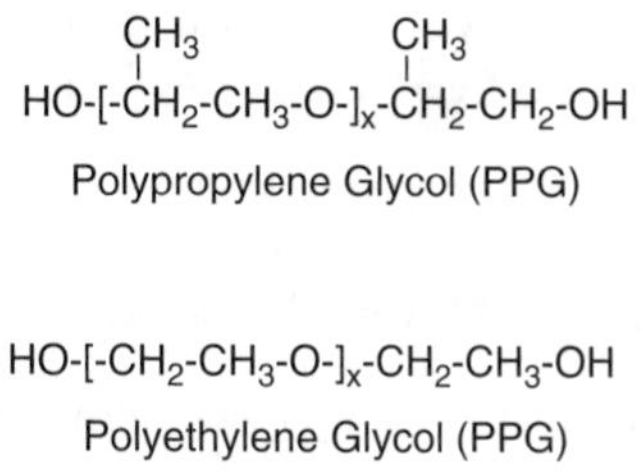

$$CH_3 \qquad CH_3$$
$$HO\text{-}[\text{-}CH_2\text{-}CH_3\text{-}O\text{-}]_x\text{-}CH_2\text{-}CH_2\text{-}OH$$
Polypropylene Glycol (PPG)

$$HO\text{-}[\text{-}CH_2\text{-}CH_3\text{-}O\text{-}]_x\text{-}CH_2\text{-}CH_3\text{-}OH$$
Polyethylene Glycol (PPG)

FIGURE 2.5 Polyether polyols.

Readers, however, should not be prejudiced by these comments. The important consideration is the condensation of any polyalcohol with an isocyanate. Inasmuch as the polyalcohol is the compound that gives us the opportunity to produce a chemically active polymer, a researcher should not be limited by the history of polyurethanes that was guided by the need for a physically strong polymer system. In any case, discussing polyether polyols is a suitable starting point.

Polyethers are typically products of base-catalyzed reactions of the oxides of simple alkenes. More often than not, ethylene oxides or propylene oxides and block copolymers of the oxides are used. A polypropylene oxide-based polymer is built and then capped with polyethylene oxides. An interesting aspect of this chemistry is the use of initiators. For instance, if a small amount of a trifunctional alcohol is added to the reactor, the alkylene oxide chains grow from the three alcohol end groups of the initiator. Suitable initiators are trimethylol propane, glycerol or 1,2,6 hexanetriol. The initiator is critical if one is to make a polyether foam for reasons that we will discuss shortly.

While the use of these polyethers is widespread, the goal of discussion is to create a specialty chemical. Propylene- and ethylene-based polyols are produced for physical reasons and will serve as the backbone. Researchers should note, however, that the scope of polyethers and polyesters is much broader when they are willing to sacrifice some physical strength to gain a chemical advantage. To illustrate, we cite a particularly interesting example. Castor oil was a common polyol for the production of polyurethanes. It was replaced by less expensive and more predictable polyols in commercial production. Readers should be aware that mixed polyols can be used to advantage.

Returning to conventional technologies, the use of polyethers in polyurethanes is relatively recent. The first reports were based on experiments with copolymers of ethylene oxide (EO) and propylene oxide (PO).[10]

This discussion of polyols is important because polyols provide us with opportunities for chemical designs. In the first chapter, we postulated that polymerized polyethylene glycol could be used as a solvent extraction medium. In this sense, the isocyanate is simply a means to an end. If other immobilization techniques were as cost effective and simple, they would serve as well. In short, the polyol is our chief design tool.

We have described the basic building blocks of a polyurethane. A host of other components can be and are used. They include processing aids such as catalysts, surfactants, and emulsifiers. We will mention them briefly. For a more complete discussion of those topics we suggest texts written by Oertel[11] and Saunders and Frisch.[12]

BASIC POLYURETHANE REACTION

A polyurethane is formed by reacting a hydroxyl-terminated polyether or polyester with an isocyanate. An example in commercial practice is the reaction of toluene diisocyanate and polypropylene glycol (PPG) to produce one of the most common forms of polyurethane (see Figure 2.6).

A number of issues related to this reaction should be discussed. First, a polymer is rarely isolated in this form. In the early 1950s a technology was developed that has since come to be known as the "one-shot" process. While the technique certainly produces a capped polyol, it immediately reacts further to achieve its ultimate form (Figure 2.6, bottom). You will notice that the capped polyol still has isocyanate functionalities as end groups. Regardless of the process, these end groups must continue to react (by the addition of water and/or a catalyst) to complete the process. While this reaction produces one of the most commonly constructed polyurethanes, it is rarely isolated as an end product.

The reaction shown in Figure 2.6 produces what is referred to as a prepolymer via the production method of choice before the one-shot process was developed. Prepolymers are still commonly used. Small molding operations, elastomers, and hydrophilic polyurethanes involve production of prepolymers.

If one were to design a new polyurethane, the prepolymer method is doubtless the method that would be used. *Prepolymer* is a term of art that designates an intermediate process in the production of polyurethanes as we know them. Prepolymers are quite easy to produce in a laboratory. The isocyanate is slowly added to

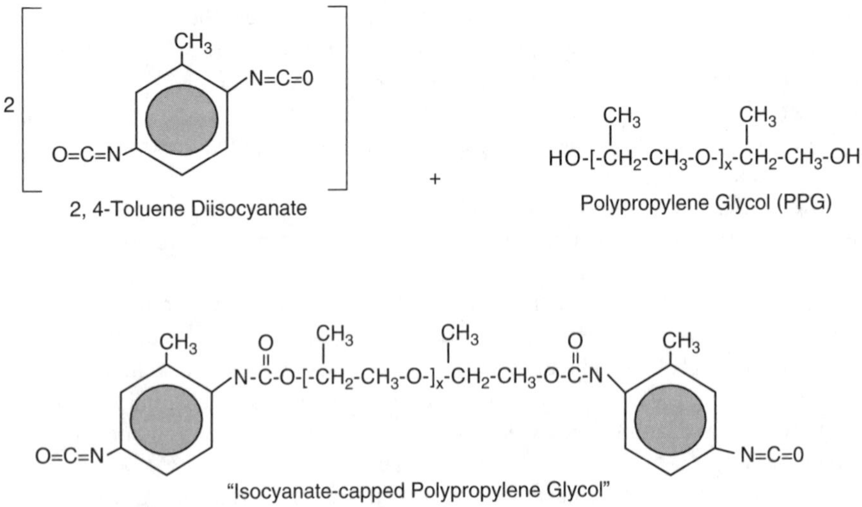

FIGURE 2.6 Polyurethane reaction for producing prepolymer.

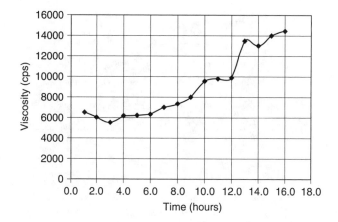

FIGURE 2.7 Increase in viscosity of IPDI prepolymer as function of reaction time at 125°C.

the polyol with continuous stirring at ambient temperature; the temperature should be monitored. The amount of isocyanate required is calculated from the hydroxyl content of the polyol.

In the reaction shown in Figure 2.6, assume that we are beginning with a PPG with a molecular weight of 1000. There are two hydroxyls (the equivalent weight of the PPG is 500). Two moles of TDI per mole of PPG are required. For 1 kg of PPG, therefore, 542 grams of TDI are required. Excess is usually added to maintain a reasonable reaction rate. Very often, if foam is the desired final product, excess isocyanate is intentionally added to improve the foam characteristics. In that case, the prepolymer is designated as a quasi- or semiprepolymer.

The reaction conditions depend on the polyol and the isocyanate, but typically the isocyanate and polyol are mixed at low temperature and then heated to 65 to 125°C. An exotherm is evident, and the viscosity slowly builds as the reaction proceeds.

In the next experiment, a 1.05 molar excess of isopherone diisocyanate (IPDI) was dissolved in polyethylene glycol (ca. 6000 MW, hydroxyl = 3, nominal) and heated to 125°C under nitrogen. The viscosity at 25°C and the residual NCO (short-hand for isocyanate) content were determined. Figure 2.7 shows the increase in viscosity over time. The NCO concentration, a measure of the extent of the reaction, was determined by a method discussed in the next chapter.

As its name implies, a prepolymer is an intermediate step to a usable polymer. It is reacted with catalysts to complete the formation of a true polyurethane via several methods. The first is based on the fact that isocyanate groups will, in time, react with one another. Thus even the most carefully controlled prepolymer has a shelf life. Water, acid, and temperature must all be kept to a minimum to extend the useful life of a prepolymer.

Normally, however, you will want to continue the reaction in a controlled manner. The conduct of that reaction will depend on the composition of the prepolymer and the intent of the device under study. If a prepolymer is hydrophilic (the polyol being polyethylene glycol), curing might be done by the addition of water. In that case,

1.2
1
0.8
0.6
0.4
0.2
0
?
0.0 0.2 0.4 0.6 8.0 10.0 12.0 14.0 16.0 18.0
Time (hours)

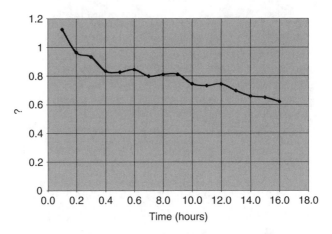

FIGURE 2.8 Decrease in isocyanate concentration of IPDI prepolymer as a function of reaction time at 125°C.

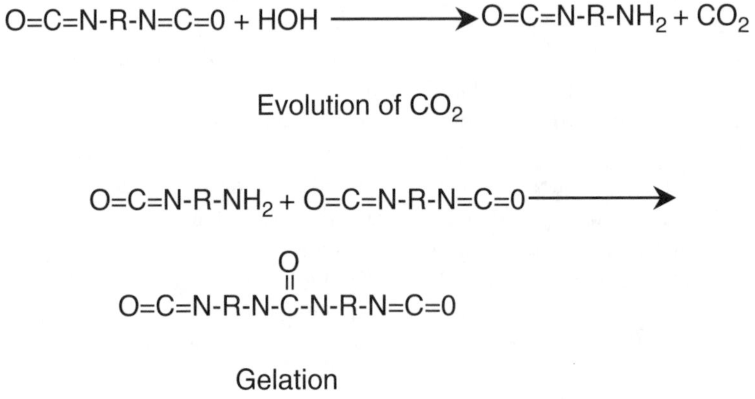

O=C=N-R-N=C=0 + HOH → O=C=N-R-NH2 + CO2
Evolution of CO2
O=C=N-R-NH2 + O=C=N-R-N=C=0 →
O=C=N-R-N-C(=O)-N-R-N=C=0
Gelation

FIGURE 2.9 Reaction of a prepolymer with water.

the chemistry in Figure 2.9 would be applicable. We will discuss polymer design in the next chapter.

The nature of a hydrophilic prepolymer permits the addition of large amounts of water. The isocyanate reacts with the water to abstract CO_2. The amine that also results from the reaction then reacts with an isocyanate group to produce a urea linkage. The reaction continues until the water or the isocyanates are consumed. If provisions are made to trap the CO_2 in the mass, a foam is produced. If such provisions are not made, the CO_2 will bubble away, leaving behind a low gel-strength hydrogel. Careful examination of the resultant molecule might cause one to rename it a polyurethane/polyurea.

If a prepolymer is made from hydrophobic polyols (as shown in Figure 2.6), the prepolymer is mixed with additional amounts of polyol to complete the reaction (Figure 2.10). Most commercially available prepolymers are based on this technique.

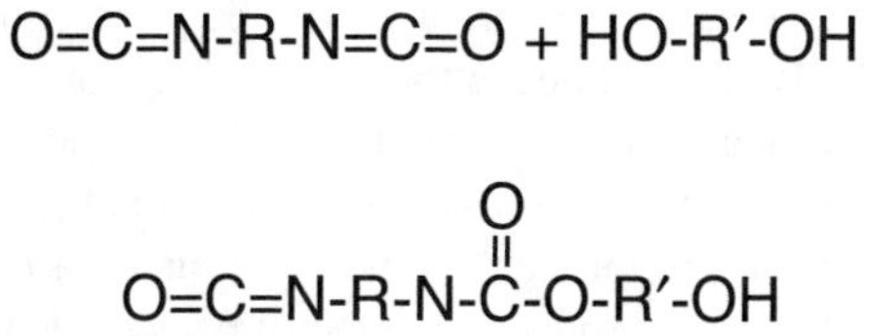

$$O=C=N-R-N=C=O + HO-R'-OH$$

$$O=C=N-R-N-\overset{\overset{\displaystyle O}{\|}}{C}-O-R'-OH$$

FIGURE 2.10 Reaction of a prepolymer with a polyol.

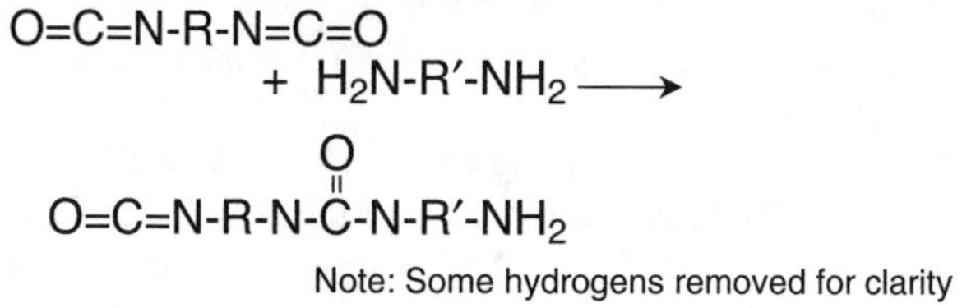

$$O=C=N-R-N=C=O$$
$$+ \ H_2N-R'-NH_2 \longrightarrow$$

$$O=C=N-R-N-\overset{\overset{\displaystyle O}{\|}}{C}-N-R'-NH_2$$

Note: Some hydrogens removed for clarity

FIGURE 2.11 Reaction of a prepolymer with an amine.

They are typically sold as two-part systems. By convention, the isocyanate prepolymer is designated Part A while the polyol, selected catalysts, and perhaps even water comprise Part B.

Comparing Figure 2.9 and Figure 2.10 reveals that the linkages between the polyol and the isocyanate are the traditional urethane linkages. You will notice also that CO_2 is not liberated by this reaction. If a foam is to be made, a blowing agent or small amount of water is added. The last reaction of a prepolymer involves a diamine or dicarboxylic acids (Figure 2.11). The foamer is a more common approach, and numerous commercial urethane-grade diamines are available for this purpose.

In this reaction, as with the polyols, no gas evolves. Thus, if a foam is to be made, it must be produced via the addition of a so-called blowing agent or by including a small amount of water (0.2 parts per part of prepolymer). As one might expect, the use of an amine curing agent results in a multiplicity of urea linkages. This reaction is typically very fast. By including large and complex diamines, satisfactory elastomers are produced.

We have alluded to an aspect of prepolymer construction related to cross-linking and not depicted in Figure 2.6. As a rule, if one is to make a foam, a small amount of three-dimensional character is required. The concept is that when gas is generated in the reacting prepolymer, either by the use of a blowing agent or by the inclusion of water (to produce a "water-blown foam"), provision must be made to trap the gas in the matrix. If a linear prepolymer is produced, it will increase in viscosity as polymerization proceeds, but the internal pressures will generally exceed the strength of the polymerizing mass. The gases will escape, leaving an elastomer behind. If the prepolymer were based on a hydrophilic polyol, the mass would be designated a hydrogel that would not have much physical strength and therefore would be of little use. This is rarely a clean process, however. Some gases are trapped so that the mass is of unpredictable density.

In most cases, linear prepolymers are used for the production of elastomers. Gas evolution via dry polyols or diamines is therefore not encouraged. When a foam is required, however, a slight adjustment is made in the construction of the prepolymer. We referred to it earlier when we discussed polyfunctional polyols. By including a trifunctional alcohol in the prepolymer recipe or by using a polyol from a trifunctional initiator (e.g., trimethylol propane), a three-dimensional character is introduced into the prepolymer.

In designing a prepolymer for a specific use, this technique provides for more control of the chemistry. For elastomers, for instance, including some three-dimensional character changes the physical properties of the polymer. More cross-linking lowers elongation and increases strength. Cross-linking is not necessary for elastomers, but it is required for foams.

Two methods are available for adding cross-linking. A small amount of glycerol or trimethylol propane (3% based on the mass of the polyol) can be added and then processed as described above (Figure 2.5). Alternatively, the prepolymer can be processed as noted above. A certain amount of internal cross-linking will occur when it is heated above 125°C. Polyols available from manufacturers are already cross-linked. They can be purchased with hydroxyl contents as high as 4 or 5. The higher the hydroxyl number, the more cross-linking.

With both techniques, the prepolymers are significantly changed by cross-linking. Because of the cross-linking, as the polymer reacts, the system gels. Without the cross-linking it would simply have increased in viscosity. Because of the three-dimensional character, the CO_2 becomes trapped in the matrix and this begins the process of making foam. The change from a true liquid with a measurable viscosity (without cross-linking) to a gel of infinite viscosity is the key to manufacturing foam. The point at which the gel structure is created is called "cream time."

As the mass increases in volume due to the production of gas, a competition between the gas pressure and the gel strength of the polymer ensues. If the former is higher than the latter, a point will be reached at which the foam will collapse. The easiest way to demonstrate this is by increasing the temperature. The reason relates to the activation energies of the two reactions (polymerization and abstraction of CO_2). If the temperature is increased, the rate of gas evolution increases faster than the polymerization or gelation.

Lowering the temperature produces the opposite effect. The polymerization reaction depends on abstraction first; the balance of the reaction favors polymerization, and a high-density foam is produced. This is a complicated effect when incorporating a temperature-sensitive component in the foam. We will discuss the addition of living cells into a foam sample later. Controlling the temperature is an important consideration. Unfortunately, foam quality is sacrificed because the reaction must be conducted at reduced temperatures.

If the temperature is carefully controlled, the mass will increase in volume until the mass is strong enough or the evolution of gas is low enough to prevent further expansion. This is referred to as "rise time." The molecular weight is still increasing at this point. As it increases further, it passes through a point at which it no longer has adhesive character. This is known as "tack-free time." If the purpose of polyurethane is to serve as an adhesive, chain terminators are added to stop the reaction while the mixture is still adhesive. We will discuss this further when we review structure–property relationships. The reaction continues with a gradual increase in molecular weight and the associated increase in physical strength. The reaction from the first time the prepolymer comes into contact with water lasts about 30 minutes.

We have focused our attention on the prepolymer method for polyurethane development because we feel that it offers the researcher the greatest control of the molecule. We hope to encourage the scientific community to investigate other nonclassical polyurethane tools. If the purpose of a device is purely physical, the large polyurethane manufacturers and chemists are the best resources for expertise. If, however, the intent is to experiment with a new polymerization technique for a particular medical or environmental application, the researcher must be able to assemble the component parts along lines with which the polyurethane industry may not be familiar.

Hydrophilic polyurethane is a good example. Researchers at W.R. Grace & Co. needed a hydrogel material of significant strength to be used as a printing plate. The material also had to be hydrophilic, however, and for a reason lost to history, ethylene glycol was chosen as the backbone. For many of the same reasons discussed in the first chapter, the isocyanate reaction was chosen. The researchers performed a polycondensation using a polyol and an isocyanate. The technique led to the development of the Hypol business at Dow Chemical. The material never was used as a printing plate.

It is helpful to understand the polyurethane industry as it exists today starting with the process most commonly practiced. The basic reaction of an isocyanate and a polyol is still used, but large manufacturers do not isolate the compound as an intermediate.

The "one-shot" process was developed in the 1950s. It is a technique by which the polyol and the isocyanate are mixed together. Without help, however, the reaction would be too slow to be useful. For this reason, the key to the process is the development of a formulation that accelerates the isocyanate reaction. We will discuss the design of a formulation to meet a specific end use in the next chapter. To complete our discussion of the processes by which polyurethane foams are made, consider the formulation for a conventional foam used for furniture cushions (Table 2.3).

TABLE 2.3
One-Shot Formulation for
Producing Flexible Foam

Component	Parts
TDI	51.6
Polyol	100
Catalyst	0.5
Water	4
Surfactant	1

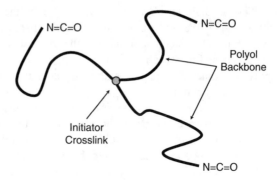

FIGURE 2.12 Cross-linked prepolymer.

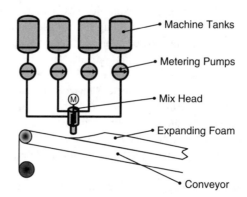

FIGURE 2.13 One-shot process for production of polyurethane foam.

Compare the complexity of this formulation with the reaction shown in Figure 2.6, which shows the production of a prepolymer as an isolated intermediate. All these reactions occur within the 30 minutes it takes to translate the mixed ingredients into a stable foam suitable for furniture cushions. Figure 2.13 shows in schematic form the equipment in which this reaction is conducted. The key to the process is the mixhead. The two basic techniques are designated as high pressure and low pressure.

High-Pressure Technique — The components of the formulation are pumped at high pressure (1500 to 3000 pounds/square inch) into an impingement mixer. As the name implies, the mixing is achieved by collisions of the ingredients at high velocity.

Low-Pressure Technique — Mixing of the components is achieved by a mechanical mixer in the mixhead. More often than not, the mixing is done by rotating pins within a metal housing. Usually referred to as pin mixers, these devices spin at 2000 or more revolutions per minute.

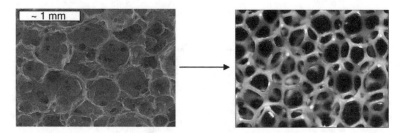

FIGURE 2.14 Conversion of open-cell foam to reticulated foam.

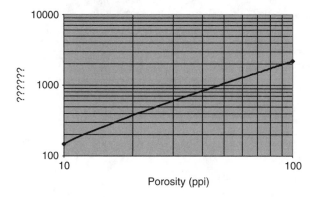

FIGURE 2.15 Surface area of reticulated foam as a function of porosity.

RETICULATION

We noted earlier the benefits of a specialty grade of polyurethane known as reticulated foam. Figure 2.14 shows a typical open-cell foam produced by the one-shot process. The right side of the figure shows how the foam might look if reticulated. A visual examination of the two foams in Figure 2.14 would lead to the conclusion that the reticulated foam would have a lower pressure (or higher flow rate), i.e., it would have less resistance to flow. As we build our case for using the chemistry of polyurethane to effect changes in the fluids passing through it, it seems clear that a reticulated foam would present significant engineering advantages. In the next chapter we will discuss an analytical technique known as "air flow-through" — a measure of the pressure needed to pump air through a standard foam (or flow rate at constant pressure).

While the primary reason for reticulation is to improve flow-through characteristics, it provides a further benefit by making the surface available to fluids passing through. The technology also produces a remarkable degree of uniformity in cell size. This contributes to the predictability of both flow and surface characteristics. If the surface is activated in some way, it is easy to see why this aspect of reticulation could be beneficial in designing functional devices. Table 2.4 and Table 2.5 show typical physical properties of commercially available reticulated foams.[13]

TABLE 2.4
Properties of Reticulated Foams

Porosity (pores/inch)	Tensile Strength (psi)	Elongation (%)	Compression Deflection	
			25%	65%
10	20	315	0.48	0.72
20	25	320	0.42	0.67
30	25	320	0.40	0.65
45	28	340	0.40	0.65
60	33	402	0.40	0.65
80	35	415	0.40	0.65
100	35	415	0.49	0.65

TABLE 2.5
Nominal Pore Size Ranges

Porosity Grade (pores/inch)	Minimum Porosity	Maximum Porosity
100	80	110
80	70	90
60	55	65
45	40	50
30	25	35
25	20	30
20	15	25
10	8	15

If we imagine that the surface has an effect on fluids that pass through it, the kinetics of the action in part serve as a function of the surface area. Again, reticulated foam presents some advantages.

The big question is whether the seemingly ideal properties of a reticulated foam will be maintained when we start to change the chemistry (for cell adhesion, extraction, etc.). Changes in the polyol or isocyanate will inevitably affect its physical properties. A balance of chemical activity would have to be established. In many cases, this balance will degrade the desirable properties. An answer is the recent development of a composite of a chemistry designed according to desirable chemical features grafted to a reticulated scaffold.[14] Such a composite was developed and patented and it will be cited as an example for several applications.

The chemistry function is produced by the prepolymer technique. Examples will follow in later chapters. The prepolymer is mixed with a reactant (polyol and/or water) and immediately pressed into a reticulated foam using a nip roller arrangement as shown in Figure 2.16.

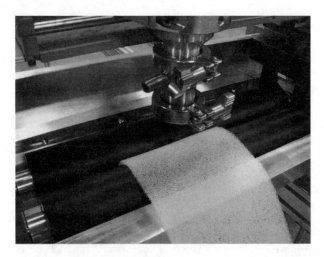

FIGURE 2.16 Nip rollers for forcing prepolymer emulsion into reticulated polyurethane foam web.

A full review of the formulations and engineering is impossible, but it is hoped that this introduction will be useful in providing a context within which we can begin to design specialty grades of polyurethane. In the past several sections, we focused our attention on the physical aspects of polyurethane. This will be extremely useful information. While we may change the molecules, we will have to stay within the context of the known processing techniques, and these sections are intended to provide a foundation on which to build. The polyurethanes we will design will be governed by the chemical effects of the polymers. By way of example, we will discuss the attachment of living cells to a polyurethane. This requires that we build a complex polyol, which necessarily changes the processing techniques and the properties of the final product. If this polymer system is to have a physical structure, however, we cannot stray far from conventional polyurethane technology. In extreme adjustments of the polyurethane, a composite approach as mentioned above might be advisable.

HISTORY AND CURRENT STATUS OF POLYURETHANES

We will conclude this chapter with a commercial perspective of polyurethanes. It will be useful to see the research that has taken place in this field over the past 70 years. The intent was to devise a polymer system with physical characteristics of flexibility, tensile strengths, and other factors that could be applied in the development of useful devices. We will discuss chemical features in subsequent chapters.

The first modern polyurethanes were developed in Germany in the late 1930s as attempts to produce polymer systems that would be useful as substitutes for natural latex rubber in automobile and truck tires, clearly in preparation for war.

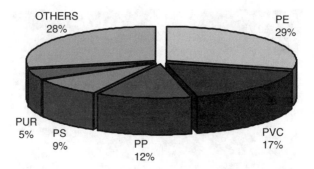

FIGURE 2.17 Production of polymers by percent. PE = polyethylene. PVC = polyvinyl chloride. PP = polypropylene. PS = polystyrene. PUR = polyurethane.

Work continued during the war and patents describing the production of flexible polyurethane foams were issued. The first commercial products based on polyester polyol technologies were not available until the 1950s. The introduction of polyether polyols in the late 1950s led to great improvements in comfort, durability, and resistance to hydrolysis. Progress continued with the development of new catalysts and surfactants. Certain process improvements produced significant cost reductions as well as improvements in foam quality. Significant quantities of polyurethane foam and thermoplastics became available in the 1950s in Europe and North America. While most commercial interest focused on foam applications, the use of polyurethane technology for so-called elastomeric forms progressed steadily. These high density materials currently play and will continue to play increasingly important roles in commercial applications. The polyurethane market today amounts to about 5% of the total polymer market.[15] In 1990, 105 million tons of polymers were sold in the volumes shown in Figure 2.17.

The impact of polyurethanes has been global but is mostly concentrated in the developed world. The availability of the raw materials and the equipment to manufacture and process the materials permits small operations to exist economically. High-quality foams and elastomers can be manufactured in small facilities without large capital expenditures. The worldwide consumption of polyurethanes has therefore increased steadily since the 1950s. Current consumption by region is shown in Figure 2.18.

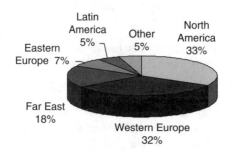

FIGURE 2.18 Polyurethane production capacity by region.

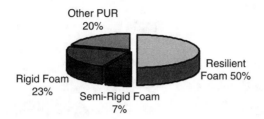

FIGURE 2.19 Polyurethane usage by type.

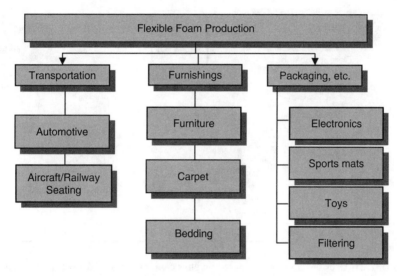

FIGURE 2.20 Uses of flexible foam.

One of the great benefits of polyurethane is versatility. With only slight changes in chemistry, one can make products ranging from soft furniture cushions to automobile bumpers and infinite numbers of other products. Depending on the application, a polyurethane chemist can vary density and stiffness to achieve acceptable product performance. The chemistry is in fact much more versatile than is required. Figure 2.19 covers soft foams, rigid foams, and other polyurethanes. We will provide more details later in this chapter, particularly as to how the independent properties of density and stiffness relate to end uses.

Rigid foams are used for structural and insulation uses while the flexible materials are used for a vast variety of applications as seen in Figure 2.20. The versatility of polyurethane positions the product as unique in the polymer world because of the breadth of applications. As we will show, small changes in chemistry can achieve a broad range of physical properties. This statement emphasizes the physical properties and serves as a testament, however, to the lack of chemical interest. It is supported by a description of the independent variables of density and stiffness and the range of products based on the primary attributes of polyurethanes. See Figure 2.21.

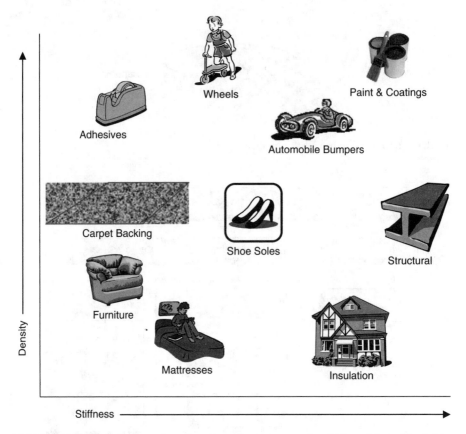

Density

Stiffness

FIGURE 2.21 Uses of polyurethane based on physical properties. (Adapted from Wood, G., *The ICI Polyurethanes Book*.[16])

Before we leave this descriptive chapter and explore the simple molecular chemistry that allows the diversity of applications, we must discuss a departure from traditional applications for polyurethanes. Readers will notice that the above discussion of applications focuses exclusively on physical properties; we have painted a picture of polyurethanes that indicates the only reason to use them is their tensile or compressive strength. Based on Figure 2.21, polyurethane is capable of holding things together, protecting them, or fostering comfort.

One grade of polyurethane used for its chemistry has come to be known as hydrophilic polyurethane (HPUR). Invented in the early 1970s in the research laboratories of W.R. Grace & Co. in Columbia, MD, it holds a unique position in the world of polyurethanes in that it is valuable because of its chemical nature although it is an insignificant product from the perspective of its volumetric. Probably no more than 50 million pounds are produced worldwide. Unlike common polyurethanes, however, HPUR is used as a drug delivery system, in wound care products, for fragrance and soap-delivery systems, as a component in a host of medical devices, as a semipermeable membrane in protective fabrics, and as specialized growing media for high-value plants. What makes HPUR unique is that it is never used for structural reasons to provide strength and resilience. It is used because its compatibility with water-based systems is essential for the proper functioning of the devices and compounds in which it is contained.

The size of the HPUR market belies its importance, at least as a surrogate for the use of polyurethane as a chemical. We stress the fact that HPUR is used because the constituent parts of its polymer backbone affect materials with which they come into contact — it is not used to provide strength. Its simplest model absorbs water, but in more complex associations, microorganisms can colonize because of a property known as biocompatibility. In fact, all polyurethanes exert effects on materials in their environments. This book is intended to illustrate that and change the perception of all polyurethanes from simple polymers to physical–chemical devices with absorptive, extractive, and biocompatibility properties.

3 Structure–Property Relationships

In the first two chapters, we introduced polyurethane, not from the traditional chemistry view, but from the perspective of researchers desirous of a system to solve problems not typically addressed directly by polyurethanes. It has rarely been the intent of researchers in polyurethane to develop anything but a physical polymer system. Increasing tensile strength and controlling rigidity, flammability, softness, and hardness guided formulation development.

Our intention is to broaden the design parameters of new polyurethanes that include chemical features such as cell binding and extraction. To that end, we have presented a primer on modern polyurethane design and discussed the current library of polyols, isocyanates and, to a degree, additives. Remembering that almost without exception the current library was assembled to solve physical problems, our library must expand to concepts in physical chemistry. The exception to this general description of the course of typical polyurethane research has been the development of so-called hydrophilic polyurethanes. Both systems we will discuss were developed specifically to take advantage of the chemistry of polyurethanes. The intent of these technologies is consistent with the philosophy of this book: production of specialty chemicals.

It is useful to present a metaphor. The Food and Drug Administration (FDA) differentiates medical devices from pharmaceutical products by their primary purposes. Based on FDA guidelines, a device meets its intended purpose by physical means. An example is a wound dressing. A pharmaceutical product meets its purpose via chemical means, for example, a drug. To carry the metaphor still further, the FDA recently added a new organization that focuses on "combination products" that are analogous to the composites we will discuss, i.e., products with physical and chemical aspects. The development of the biofilter for the remediation of polluted groundwater is one example we will discuss. A biofilter is physical in the sense that the scaffold of the composite provides strength, high void volume, high surface area, and low pressure drop. A grafted hydrophilic surface provides reservoir capacity for storage and buffering of nutrients and a biocompatible surface for cell growth.

We will discuss several examples in subsequent chapters, but for now, we continue our primer on basic polyurethane technology in order to develop further skills required to build polyurethanes. This chapter covers traditional polyurethanes with physical design features in an exploration of how polyurethanes are designed. We will also discuss most of the parameters that define the current state of the art.

One of the great strengths of polyurethane chemistry is the degree to which one can change its design features with small changes in chemistry. In the field of hydrophilics, the same capital equipment and reaction parameters can be used, but small changes in the polyol and in the degree of cross-linking can make a cosmetic applicator sponge or hydrogel suitable for use with contact lenses. On the hydro-phobic side, the processes to develop seat cushions and automobile bumpers are remarkably similar. We will discuss the processes later in this chapter. However, it is not the intent of this book to teach readers to manufacture such products. Our intent is to provide conventional polyurethane tools that can be used by readers to initiate the process of designing physical aspects of new devices. It is assumed that readers can identify the design requirements and separate the physical from the chemical issues. This chapter will show how physical needs can be met through standard polyurethane design disciplines.

ANALYSIS OF POLYURETHANES AND PRECURSORS

While the object of this chapter is to introduce readers to the science of polyurethane design, it is best to begin with definitions of terms. When physical requirements are to be addressed, quality system disciplines dictate that a system of analysis be established to monitor compliance with the design. For instance, if a device is to allow the flow of water through a foam, a system that can be validated to indicate flow rate or pressure drop must be established.

In the world of polyurethanes, standard methods describe the properties of the materials. While the conventional methods of analysis are necessary for compliance with the physical design of a system, our task must include chemical tests, and they must be developed on an *ad hoc* basis. Nevertheless, conventional methods for analyzing a polyurethane are very helpful and are reviewed here. Our focus will be on the effects of the polyurethane chemistry on fluids with which it comes into contact. Because this is different from the typical use pattern, some of the standard methods used to produce seat cushions, for example, will not apply. We will discuss several methods but not in the traditional order.

The primary sources for methods of polyurethane foam analysis are the American Society for Testing and Materials (ASTM) and the International Organization for Standardization (ISO). Each organization has set specific standards for each of the several forms of polyurethanes, i.e., elastomers, coatings, and foams. Of most interest to us is the testing of foams, also known as *cellular polymeric materials*. The reason relates to our intent to focus on the effects of polyurethanes on fluids (gases and liquids) with which they come into contact. As noted, foam materials, particularly open-cell foams and most particularly reticulated foams, combine high surface area, low pressure drop, and high void volume. These properties and others make retic-ulated foams useful in the contexts discussed in this book. That is not to say that elastomers and other forms of polyurethanes are not useful. As noted, their chem-istries are nearly identical, and a particular chemical feature of one is maintained in the other forms. This concept is most obvious when one considers the composite materials we have touched upon. Nevertheless, foams provide convenient platforms on which to build devices that take advantage of chemical activities.

In keeping with the scope of this book, we will focus our attention on those procedures that most appropriately define a flow-through system. ASTM D 3574-95 lists eight tests for defining a foam sample:

- Density
- Indentation force deflection (IFD)
- Compression force deflection (CFD)
- Compression set
- Tensile strength
- Tear resistance
- Air flow
- Resilience

Some important aspects of foam are not included in this list. We mentioned void volume earlier. This parameter is related to density by a comparison of bulk density to absolute density. We will discuss each of these measures of foam quality and relate their applicability to the theme of this book. Later in the chapter we will show how formulation and processing techniques are used to adjust and control the most important parameters.

DENSITY

Density is an important property because, all other things being constant, it is colinear with the compressive and tensile strength of the foam. It is interesting to note that it is not colinear with pore size. By visual examination, one would assume that a 100-ppi foam would have a higher density than a 10-ppi foam. In fact, they can be made to have the same density because the bars and struts that form the foam matrix become thinner as pore size gets smaller. Table 3.1 lists the porosities and densities of several commercially available reticulated foams.[17]

The void volume can be determined by displacement in water. This is problematic when investigating a hydrophilic polymer or when accurate measurements are required. A good estimate of void volume can be obtained by comparing the absolute density of the polymer with the bulk density determined via ASTM D3574-95. For instance, let us assume that isocyanate and the polyol have specific gravities of 1.0. If a polyurethane is made of this combination, the absolute density of the polymer

TABLE 3.1
Comparison of Porosities and Densities
of Reticulated Foams

Porosity (ppi)	Density (lb/ft³)
10	1.9
20	1.9
30	1.9
45	1.9
60	1.9
80	1.9
100	1.9

would also be 62.4 lb/ft³ (specific gravity (SG) = 1.0). If the bulk density is measured according to the ASTM method and found to be 2 lb/ft³, a cubic foot of the foam would be 2/62.4 or 3.2% of a cubic foot. The remainder (96.8%) would represent the void volume.

The void volume will become an important factor when we begin to discuss reaction rates in the environmental sections of this book. Void volume also exerts an effect on the inhibition of flow through the foam, not to mention turbulence and other mass transport phenomena.

COMPRESSION

Compression methods are most often used to indicate softness and they can determine the thickness of a sample of foam under an applied force or vice versa. While *stiffness* and *compressibility* are terms used interchangeably, they are different phenomena. Stiffness is resistance to bending, while compressibility is, as the name implies, the resistance to squeezing. While we must be aware of the semantics, in practical terms the method for changing both factors is the same. If we increase compressive strength, we also increase stiffness.

Two techniques allow determination of the compressibility of foam. The methods are related and the application dictates which method better defines the quality of the product. For example, for furniture cushions, indentation force deflection (IFD) is a more common test technique than compression force deflection. Both methods determine the amount of force required to compress a foam to a percentage of its thickness. In the IFD test, the plunger that compresses the foam is smaller than the foam sample. This presumably correlates to a person sitting in the center of a chair cushion.

After a process by which the sample is preconditioned, a plunger is pushed into a foam sample as illustrated in Figure 3.1. A load cell attached to a plunger records the force required. The area of the circular foot of the plunger is 325 cm². The force required to compress the foam to a given percentage of its thickness or the percentage of compression for a given force is determined. Most commonly, the former is reported. The forces at 25% compression and 65% compression characterize the features of the foam most important to furniture manufacturers. The "support factor"

Force Applied

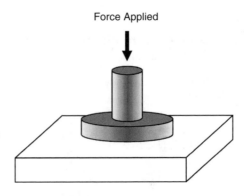

FIGURE 3.1 Indentation force deflection.

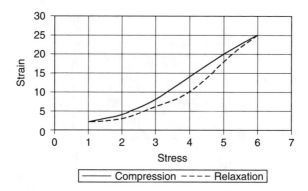

FIGURE 3.2 Hysteresis of polyurethane foam under compression.

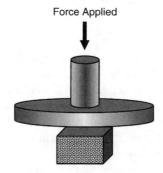

FIGURE 3.3 Compression force deflection.

or "comfort factor" is defined as the ratio of the 65% to 25% IFD. After compression, as the plunger is drawn back, the force exerted on the plunger by the relaxing foam is not identical to the force required to compress. This effect is known as *hysteresis*. Figure 3.2 shows the hysteresis effect. In many of the applications we will discuss, compression and therefore hysteresis may become factors.

Another measure of compression is actually more appropriate to our applications. The method of analyzing compression and relaxation is the same and only the designs of the feet are different. Instead of occupying a fraction of the surface area of the foam, the foot is bigger than the sample. The remainder of the test is the same. In some of our applications, we will apply a force to the surface of a foam column by passing a fluid through it. Inasmuch as the force will be assumed to be uniform across the foam, CFD would appear to be more consistent with how the foam is to be used. Figure 3.3 presents a schematic of the technique. The force required to compress the foam to 50% of its original thickness is reported.

It is important to note that the compressive strength is controlled by the chemistry to some degree, of course, but other than that, the phenomena are physical. Cell structure, density, and other physical factors control properties of the material. It must be noted, therefore, that with the exceptions of composites, the polyurethanes

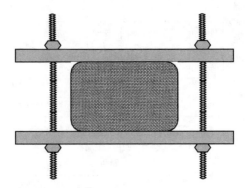

FIGURE 3.4 Compression set.

we will design based on the requirements of a proposed chemical effect will have strong impacts on the compression properties of the polymers.

Compression Set

If foam is held in a compressed state for a time, it does not return to its original thickness. This effect is referred to as compression set and could be important because the foams we will be designing will be under more or less constant compressive stresses in normal service. The analytical method involves a certain amount of compression of the foam. It is held in the compressed state for 22 hours at 70°C, then allowed to relax for 30 minutes. The thickness is measured and compared to the original thickness. See Figure 3.4.

It should be noted that the compression set test is often conducted on wet foam or in a humid atmosphere. This implies that the numbers will be different, and indeed this is the case. More often than not, polyurethane chemists work in wet or humid environments, and a procedure that uses this option is advisable.

Tensile Strength

The tensile strength of a polymer is defined as the force required to break a sample by pulling. One specific design is called a "dog bone" (Figure 3.5). One end of the device is held stationary and the other end is pulled by a device attached to a load cell. The force required to break the sample is recorded, and concurrently a measure of the elasticity of the sample is recorded as the percent elongation of the sample. As the sample is stretched, the force required to stretch the sample from 50% to 60%, for example, is reported as the modulus of elasticity. These tests are also subject to a hysteresis effect. The force/stretching curve has a yield point (or elastic limit) beyond which the sample cannot return to its original length.

It is important to note that tensile strength is controlled by density, of course, but tensile strength is otherwise a molecular phenomenon. Molecular weight, cross-linking, and other chemistry level factors affect tensile properties. We should note that, with the exception of composites, the polyurethanes we will design based

Force
Applied

FIGURE 3.5 Determination of tensile properties of polyurethane.

on the requirement for a proposed chemical effect will strongly impact the tensile strengths of the polymers.

AIR FLOW

Air flow is a measure of the resistance that a foam presents to air passing through it. Intuitively one would expect that large pore size presents less of an impediment to flow than small cells. This is indeed true and it is this method that quantifies the effect. It is obvious by now that air flow is an important property in the context of this book. Unfortunately, the ASTM tests were designed for the measurement of resistance to air flow only. Our interests focus on the flow of other fluids, specifically water, through the foam. Nevertheless, air flow represents a quick and precise way to determine the quality of a foam. If you are not sure of the difference between a reticulated foam and an open-cell foam, this test will differentiate them. Reticulated foams offer much less resistance to flow than open-cell foams.

Figure 3.6 shows the experimental setup for measuring air flow. A constant flow rate of air is passed through the sample and the pressure drop is recorded. Alternatively, the flow rate of air is determined at a given pressure drop. These data provide a good indication of the porosity of the foam and are therefore very useful for in-process quality assurance and control. However, this system falls short of defining the in-use characteristics of a foam sample, especially when very high flow rates are used or water serves as the fluid. In such cases, the action of the fluid is a compressive stress and, as might be imagined, the compression has a restrictive effect on flow.

We have developed a technique specifically designed to measure the resistance to flow and the amount of compression as a function of flow rate. The technique is used to design biofilters. As a rule, a biofilter should not allow a pressure drop in

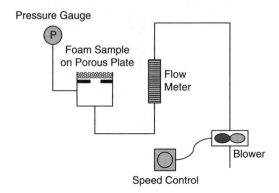

FIGURE 3.6 Apparatus for measuring air flow.

excess of 5 in. of water at the design flow rate. This factor is used to size the equipment properly.

The resistance to fluid flow is a measure of the physical structure of the foam. In order to control the flow through a foam, cell size, degree of reticulation, density, and other physical factors must be controlled. The control of these physical factors, however, is achieved through the chemistry and the process by which the foam is made. The strength of the bulk polymer is measured by the tensile test described above, but it is clear that the tensile strengths of the individual bars and struts that form the boundaries of an individual cell determine, in part, the qualities of the cells that develop. A highly branched or cross-linked polymer molecule will possess certain tensile and elongation properties that define the cells. The process is also a critical part of the fluid flow formula, mostly due to kinetic factors. As discussed above, the addition of a polyol and/or water to a prepolymer initiates reactions that produce CO_2 and cause a mass to polymerize. The juxtaposition of these two reactions defines the quality of the foam produced. Temperature is the primary factor that controls these reactions. Another factor is the emulsification of the prepolymer or isocyanate phase with the polyol or water.

In the production of a hydrophilic polyurethane, the choice of surfactant or emulsifying agent is the design tool of choice to develop a specific type of cell structure. While flow-through is not an important factor for hydrophilic polyurethanes, control of the cell structure in composites based on use provides an opportunity to develop multiples of the surface area in standard reticulated foams.

Finally, catalysts are very useful in controlling air flow. By controlling the sizes of cells or by causing the windows between cells to burst (to produce an open-cell foam), catalysts represent important tools in the effort to control air flow-through.

It is important to note that it is common in the reticulated foam industry to use air flow-through to define pore size. Manufacturers calculate average pore sizes from air flow measurements. Presumably, one would do this via a microscope either by counting pores or using an optical scanning device, but given the problems of sampling a bulk material with a microscope, it has become standard to use an empirically derived correlation between air flow and pore size. This serves as an indirect confirmation of the effect of the quality of the foam and its dynamic characteristics under flow conditions. This correlation, of course, does not apply to open-cell foams.

Other tests are cited in ASTM 3574 and other standards, but they are of less importance in the context of how we will use them. Among parameters tested are resilience, fatigue/durability, flammability, and creep. A polyurethane devised to meet our primary objectives of density, tensile strength, compression, air flow, etc., will probably lead to values for these other properties that will have to be adjusted on a systems basis.

In keeping with the general theme of this book, our task is somewhat more difficult inasmuch as we routinely must deal with the effects of water on systems. The tests described in ASTM 3527D are designed to describe the qualities of hydrophobic materials. More often than not, the products that are the subjects of this book will be in moist environments. In many cases, the products will be immersed in water. While the dry tests provide some guidance, it is clear that tensile, compressive, and water flow-through tests are needed to determine and describe the in-service properties of the materials.

Most methods can be adapted to meet the alternative use conditions that may be experienced. In the ASTM method, a preconditioning regime is suggested before compression strength is measured. In an earlier book on hydrophilic polyurethanes, we suggested soaking a foam in water for an hour, squeezing it to remove excess water, and then conducting the necessary tests. We will leave it to professional analytical chemists to develop a test capable of validation to develop a standard method.

We simply allow enough time for the polymer matrix to become saturated, as evidenced by swelling. While the polymer is hydrating, it swells to accommodate the water that hydrates the polymer backbone. When the polymer is fully hydrated, swelling stops. Mass is not a precise enough measure because one cannot account for the water on the large wetted internal surface. The surface water does not seem to have a significant effect on the tensile or compressive forces.

Once a polymer is fully saturated, the physical tests described above can be conducted with confidence. Naturally, minimizing the evaporation of water should be considered. The one exception in this new category of testing is flow of water through the foam. This is not covered in the standard but will be very important for some applications, particularly in environmental remediation. If the intent is to build a biofilter or a continuous flow enzyme reactor, we must know the hydrodynamic properties of the materials we produce. Since polyurethanes are rarely used in these environments, the flow of water even through a reticulated foam is not described by the manufacturers. Furthermore, if we are to make composites of reticulated foams, the amount of polymer grafted to the surface will have a dominating effect on the flow of water. In a later chapter, we will describe our work in this area.

A simple analytical procedure will enable us to guide the development of a number of alternative polyurethanes. The fields in which we typically work involve the interactions of polymers and water. In environmental and biotechnology applications that will play an important part in our discussions, the association with water is critical to the function of the polymer, and a test to gauge this relationship will be very useful.

In our work, we calculate the equilibrium moisture from the dry and saturated mass of samples. The method was described by Thomson.[18] We use more hydrophilic polyols, for example, hydrophilic polyurethanes and polyethylene glycol homopolymers are used. More commonly, however, we will be using polyols from the Pluronic series of block copolymers discussed in Chapter 2.

As we develop alternative polyurethanes and composites, we must be aware that an efficient system of quality validation must be developed, concurrent with progress to meet design requirements. While many of the tests described above will constitute the core of the validation procedure, additional tests are needed to fully describe the functions of the materials.

We have focused our attention thus far on the physical characteristics of a foamed polymer. If the foam that will be used is purchased other than for audit purposes, the information above should be sufficient. If, however, the foam is to be produced as part of a design, it will be necessary to supplement the analysis of the final product with procedures to ensure that the raw materials used are of consistent quality.

We have described a process by which small quantities of foam can be made by the prepolymer method. A number of methods are available to bring the variables involved in prepolymer making under control. We assume that the starting point for such processes is the acquisition of commercial isocyanates, polyols, and additives. Other than for audit purposes, we will assume they arrive with certification that they are of the so-called urethane grade. In the case of polyols, this typically means they contain less than 0.01% water and have good color. Free acids and low metal and chloride contents are important considerations for isocyanates. Manufacturers are well aware of the problems that will arise if these contaminants are not controlled and the materials cannot be used with confidence.

When the materials are converted to prepolymers, a concerted effort to control the final product must begin. Two methods are critical in this effort. The first is the determination of the −OH content of the polyol. For the polyol, the −OH content may be noted on the certificate of analysis supplied by the manufacturer, but it is useful to be able to monitor the amount of hydroxyl, especially when polyol cross-linkers are used. The stoichiometric amount of isocyanate must be known in order to control the excess. As we have shown, excess isocyanate leads to stiffness and, just as important, is an instantaneous source of CO_2 during the foaming process. Too low an isocyanate level will lead to higher-density foams. The hydroxyl number is defined as the milligrams of potassium hydroxide equivalent in 1 g polyol determined by esterifying the polyol with actetic anhydride. The −OH reacts to produce acetic acid, which is back-titrated with a standard base.[19] Alternatively, the hydroxyl vibration at 2.84 μ in the infrared range is a useful method. We combined this absorption method with a standard additions technique, adding control amounts of methanol to estimate the hydroxyl content.[20] While this technique is useful in identifying problems that may arise, the analysis of the NCO value in the prepolymer is of critical importance.

NCO has two aspects, both of which are important for controlling the polyurethane that will result. The NCO group appears in two forms. Because most polyurethanes used for foam contain excess isocyanate, they are formally called pseudo- or quasi-prepolymers. They are not differentiated in this text; we simply refer to them as prepolymers. In any case, the caps on the polyols include NCO groups that subsequently react when added to water and/or polyols. It is clear, therefore, that this influences the reaction rate and needs to be controlled. Any excess isocyanate remaining after the prepolymer preparation step has available isocyanate groups. Both isocyanate groups must be identified and differentiated in order to fully control the preparation of polyurethane.

The most reliable and convenient method to determine the total (end group and isocyanate) NCO is titration. A known amount of an amine (dibutylamine is used in the reference) is added to a weighed amount of prepolymer in solution.[21] The amine reacts with the NCO groups and the excess amine, if titrated with a standard HCl solution, to a bromothymol blue end point. The amount of NCO is calculated as milliequivalents per gram of prepolymer or mass percent of NCO. Both measures are appropriate.

The determination of the excess isocyanate is more problematic. In one method, toluene diisocyanate (TDI) is vaporized below the decomposition temperature of the prepolymer and analyzed by gas chromatography. A more precise method is useful if a liquid chromatograph is available. The NCO groups are reacted with methanol and the prepolymer is separated into its constituent parts in a size exclusion column. The methanol-capped isocyanates constitute the lowest molecular weight fraction. The isomers of TDI are differentiated by this technique.

Lastly, viscosity plays an important role in the manufacture of a polyurethane from a prepolymer. A Brookfield-type viscometer is a convenient method of analysis.

As with most areas of chemical science, the concept of analysis is open ended. The goal of fully describing a polymer involves aspects that we can only dream about being able to identify. Nevertheless, this beginning serves as a starting point to describe at least the physical aspects of the polymer. Physical strength, as described by tensile strength and elongation, has implications of how the chemistry is assembled both from a bond-cohesive aspect and in the cross-links we have developed. The compressive strength measurements, especially of foams, represent direct and indirect measures of the molecular structure to some degree but mostly reflect on the structure of the foam. Lastly, air flow speaks to the cell structure which, as we have shown, is a measure of the accessibility of the foam structure to fluids passing through it. It is clear, therefore, that we must know what factors in polyurethane chemistry permit us to control these factors.

STRUCTURE–PROPERTY ASPECTS
OF POLYURETHANE DESIGN

While we will discuss the control of the various properties presented in the last section, the composition of a device must represent an optimum of all the properties. As we have shown, the quality of a foam is achieved by a complex combination of chemical and physical effects. No unifying model combines them in a sufficiently precise way as to minimize the work involved. Thus, as we discuss ways to control compressive and tensile strength, we must be aware that these properties will affect the foaming process. The design process is made somewhat easier by using composite techniques. In this way, one can separate the physical requirements of a device from the chemical. As noted above regarding physical properties, we need tools to control tensile strength, compressive strength, and flow-through.

TENSILE STRENGTH

Regardless of the form of polyurethane, elastomer, or foam, the strength of the polymer is expressed as tensile strength. Clearly, strength is related to density reflected in the inter- and intramolecular forces that hold the polymer together under stretching stress. A polymer stretches as a stretching force is applied. The increase in force is compared to the amount of elongation and the result is the modulus of elasticity. At break, the force is reported as tensile strength and the elongation is compared to the original sample and reported as percent elongation.

We also need to discuss these factors in the resultant foam, but that discussion is complicated by density and to a degree the quality of the foam. This discussion will focus on the properties of the polymer. It is not unreasonable to assume that as we affect the polymer strength, we also affect the properties of the foam and elastomers that are produced from the polymer. Although more problematic, it also reflects on the compressive properties of the foam.

Polyurethanes follow the same rules as olefin polymers with regard to molecular weight cross-linking, intramolecular forces, crystallinity, and other factors. Molecular weight typically influences tensile strength up to some limiting value. Melting point (if the compound has one), elongation, modulus, and glass transition also reach maxima. Commonly, as the molecular weight increases, brittleness and solubility or swelling decrease.

Intermolecular forces also affect physical properties. Hydrogen bonding, van der Waals forces, and dipole moment tend to hold a polymer together under stress in a way similar to the covalent bonding of a polymer backbone. The strength of these bonds is significantly lower in energy, however. These forces can be controlled by affecting changes in the way the polymers fit together. The addition of short chain ligands prevents the coordination on which, for example, dipole–dipole interactions depend.

Crystallinity is evident in all polyurethanes and affects the physical strength of the polymers in ways similar to intermolecular forces. In fact, the factors are closely related. Crystallinity causes brittleness.

In the development of a polyurethane system, we will use the concepts of hard and soft segments to define the textual properties of a polymer. In this theory, the weight percent of so-called hard segments controls stiffness. In polyurethanes, isocyanates and crosslinking are considered hard segments. Polyethers and polyesters are soft segments, polyethers being the softer of the two. Generally, MDI-based polyurethanes are stiffer and harder than TDI-based systems. For a given isocyanate, as the molecular weight of the polyol decreases, the stiffness increases. In this sense, stiffness is a synonym for compressive strength.

Crosslinking is another important tool. It is essential to foam making because it gives the reacting mass the ability to trap CO_2. Crosslinking is important for elastomers since it controls tensile strength and elasticity.

As noted earlier, the backbone on which a polyurethane is built consists of a number of covalent bond types; the urethane bond is the least common. Each bond

TABLE 3.2
Molar Cohesive Energy of Organic Groups

Group	Cohesive Energy (kcal/mol)
–CH– (Methylene)	0.68
–O– (Ether)	1.00
–COO– (Ester)	2.90
–C_6H_4– (Aromatic)	3.90
–CONH– (Amide)	8.50
–OCONH– (Urethane)	8.74

TABLE 3.3
Relation of MDI Ratio to Properties of Resultant Elastomer

Component	Parts			
Polyester polyol	100	100	100	100
MDI	7.2	14.3	28.7	42.8
Butanediol (cross-linking agent)	9.1	11.7	16	21
Percent hard segments	14.0	20.6	29.9	38.9

Source: Saunders, J.H., and Frisch, K.C.[22]

has a cohesive energy that defines how much energy is required to tear it apart physically. This characteristic clearly has an effect on the strength of the polymer. Table 3.2 compares the cohesive energies of most of the important chemical bonds in a polyurethane.

These general comments apply to polymer science in general, but we are concerned with the specific advantages of polyurethanes. In the remainder of this chapter, we will discuss the components of a polyurethane, the process and other factors that affect the polymer, and the chemical factors that confer efficacy for environmental, medical and other applications.

We have alluded to the fact that MDI produces stiffer polymers than TDI, based on the hard-and-soft segment model of polymer design. In the experiment cited below, the concentration of MDI in a formulation was increased and the resultant polymer analyzed.

While Table 3.3 deals with elastomers, it is important to mention that the effect of MDI illustrated in the table applies to MDI foams as well. Figure 3.7 shows the effect on the tensile strength of the polyurethane of increasing amounts of MDI. The increase in hard segments increases the brittleness but does not improve the strength of the polymer, as reflected in the elongation figures. The increase also exerts positive effects on the compression of the foam as noted in the next section.

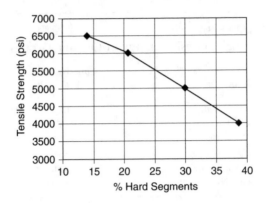

FIGURE 3.7 Effect of increase in hard segments on tensile strength of a polyurethane.

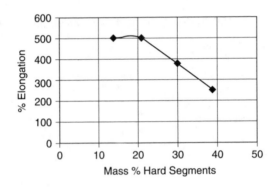

FIGURE 3.8 Effect of increase in hard segments on elongation.

Notice that the stoichiometry requires that if the polyol is kept constant in the experiment, the excess isocyanate end groups must be balanced by another polyol. In these experiments, butanediol (BTO) was used. BTO is considered a hard segment.

The changes in elongation and compression testify to the increased brittleness as seen in Figure 3.8 and Figure 3.9.

While the above discussion deals with the hard-and-soft segment concept, certain differences in the hard category contribute to tensile strength. For example, Szycher[23] reported on the effects of a number of commercially available isocyanates on the tensile strengths of polyurethane elastomers using a 500 MW poly(oxytetramethylene) glycol. It is interesting to note that the polyurethanes were made from aliphatic isocyanates. Aliphatics are less susceptible to ultraviolet weathering than the more commonly used aromatics (TDI and MDI). As one might expect, the aromatics were also differentiated by their effects on tensile strength.

While we are dealing mostly with open-cell foams, control of the degree of openness also has an effect on tensile properties. Saunders[24] reported on the control of open-cell structures by the addition of a stannous catalyst. They showed maxima

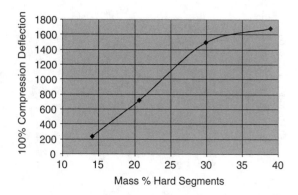

FIGURE 3.9 Effect of hard segments on compression.

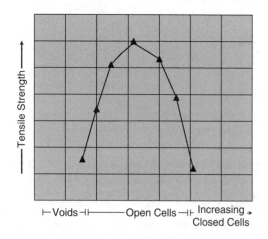

FIGURE 3.10 Effect of cell structure on tensile strength.

in their data that coincided with maxima in tensile strength. Figure 3.10 is adapted from their data.

One can develop protocols to control the tensile properties of a polyurethane by adjusting the number of cross-links in the backbone. The unit weight of polymer divided by the molar number of a cross-linking agent is referred to as the molecular weight per cross link (M_c). It is a measure of the average molecular distance between cross-links. If a cross-linking agent such as glycerol or trimethylol propane is part of a prepolymer formulation, in our experience, it is just as easy to use a mass percent cross-linker. Figure 3.11 shows the effect of cross-linking with either calculation method; it plots the slope of a tensile curve at 100% elongation as a function of M_c. This relationship allows one to use the molecular weight of the polyol to affect changes in tensile strength and elongation.

One additional factor must be covered. While not classically polyurethane-specific in nature, it is nonetheless a useful tool. The use of a filler can produce a marginal effect on tensile strength. Adding particles to a matrix reduces the tensile

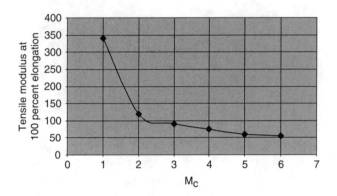

FIGURE 3.11 Relationship of cross-link density to tensile properties of a foam. (*Source:* From Saunders, J.H. and Frisch, K.C.[24]

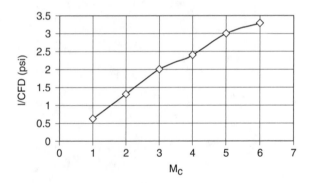

FIGURE 3.12 Effects of cross-linking density on compression.

strength while properly formulated fibers can increase. Particles can change the wear properties and density of a polymer. We often add particles to cause a foam to sink in water. The issue with fibers, apart from length and amount, is the strength of the adhesive bond between the polyurethane and the fiber. A strong bond may have significant effects on both strength and elongation.

Compressive Strength

In most cases, the factors that affect tensile properties also affect the compressive strength of a polyurethane. This is most easily depicted in a graph such as Figure 3.12 that shows compressive strengths of foams.

Cell Size and Structure

After the strength aspects of an anticipated device have been established, the design process is completed by the definition of cell structure. If the intent of the research

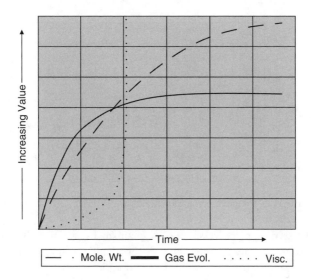

FIGURE 3.13 Juxtaposition of gas evolution gelation and molecular weight development. (Adapted from Saunders and Frisch.[25])

is to produce elastomers, this section will not be of interest, but if a project has flow-through aspects, in many ways the development of a foam can be the most critical part of a program.

Whether the "one-shot" process or prepolymer technique is used, the development of a foam involves the juxtaposition of gas generation and the development of tensile strength within the developing foam. The evolution of gas can be via the use of blowing agents or the *in situ* generation of CO_2 from the reaction of water with an isocyanate to produce a water-blown foam. In any case, the gas evolution creates an internal pressure that must be resisted by the development of a gel structure via polymerization reactions as shown in Figure 3.13.

As the evolution of gas diminishes, the strength of a polymer increases. Because of cross-linking, the mass gels (infinite viscosity) and prevents further expansion. At this point, the cells are all closed, but due in part to the gelation, the windows between the cells begin to rupture, creating the open-cell structure. A number of theories describe the process. They all mention gradual thinning of the membranes between the cells. One explanation cites a viscosity effect influenced by gravity. It is hypothesized that as the last surge of gas evolves, the high viscosity and low elastic strength do not flow fast enough to relieve the pressure. The windows then fail. If the pressure increases still further, the cells fail and are said to collapse.

We prefer an alternative model. The expansion of the foam doubtless takes place in a high-viscosity environment. Expansion of the individual cells and a corresponding increase in foam volume continue until the transition from liquid flow to gelation. At this point, the windows between the cells are stretched beyond their elastic limit and therefore burst instead of continuing to flow. Examination of a closed-cell foam shows that the windows do indeed rupture as opposed to opening through viscous flow.

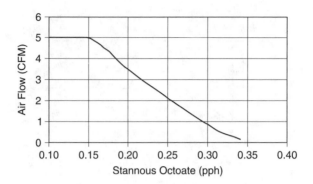

FIGURE 3.14 Effect of stannous octoate on air flow.

For whatever reason, however, the mechanism by which cell windows open would not be of much interest unless the process could be controlled. In this regard, we have two primary tools: temperature and formulation.

The reactions that lead to the evolution of gas and the increase of molecular weight (and gelation) have different activation energies. If the temperature changes, the relative rates of reaction change. Inasmuch as the development of a foam depends on the simultaneous development of molecular weight and gas evolution, a change in their rates must affect the quality of the foam. For example, if the temperature is increased above a standard value, the rate of gas evolution (which has a lower activation energy) increases faster than the molecular weight of kinetics. Thus, the internal pressure increases faster than the development of elastic strength. The result is that the foam can collapse. The opposite is also true. As the temperature is decreased, the polymerization increases faster than gas evolution, and the strength development prevents full expansion of the foam and cell rupture. We will illustrate how this has retarded the development of some hydrophilic polyurethane applications.

Temperature control is critical. While we have described this as a bulk effect, it is important to control the temperature within an expanding foam. Because the isocyanate reaction within a developing foam with water is exothermic, a temperature gradient within a bun is created. Unless provisions are made for this, the cell structure in the middle of the foam can approach adiabatic. The cell structure will be different at the elevated temperature. This is adjusted for in the production of large buns by controlling the bulk temperature and by formulation strategies that utilize a chemical method for opening. In commercial production, cell opening is controlled by catalysts or other additives including specific polyols, silicone oils, and fillers.

A common way of controlling the opening of cells is the use of a gelation catalyst. Stannous octoate (SO) is one such catalyst. Figure 3.14 shows the effects of SO on air flow. This illustration is particularly appropriate since our purpose is to build a device through which fluids will pass.

The production of an open-cell foam by the techniques described above only partially covers the polyurethanes considered most useful in the context of this book. Open-cell foams are converted to reticulated foams by a postprocessing technique. Two techniques are used in the U.S. The oldest involves immersing the foam in a

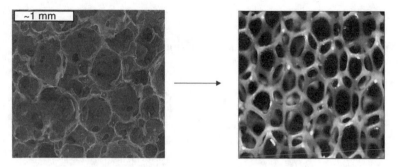

FIGURE 3.15 Transition from open-cell to reticulated foam.

strong caustic solution to hydrolyze the polyurethane. Because the windows are so think they are attached first. The other technique is permeating a foam with oxygen and hydrogen gas in a pressure chamber. A spark ignites the gases and a flame front moves through the foam, melting the windows and collapsing the matrix material into the structural members of the foam. It is interesting to note that while some minimal burning occurs, the density of the foam is not substantially affected. The effect on cell structure is dramatic, as shown in Figure 3.15.

SPECIAL CASES: HYDROPHILIC POLYURETHANE FOAMS

Special rules apply to the world of hydrophilic polyurethanes. These alternate rules are based on the fact that hydrophilic polyurethanes can and should be processed in water. Rather than emulsifying a prepolymer with a polyol, as would be done with a hydrophobic polyurethane, hydrophilics are mixed with water. While the properties of the foam are governed loosely by the guidelines described above, one has more flexibility and control of the formulation and process by which the polyurethane is made. For example, the water can serve as a heat sink to closely control the temperature of the foam; the water controls the rate of reaction.

All hydrophilics are currently processed by the prepolymer method. The emulsification of the prepolymer and water are the primary determinants of cell size. The water also serves as a heat sink to moderate the temperature of the reaction. By adjusting the temperatures of the prepolymer and the water, one can control the kinetics described above. The mass of the water limits the destructive exotherm.

With hydrophilics, control of cell opening is accomplished by the use of surfactants. Catalysts are rarely needed or used. The reason goes to the tradition of using hydrophilic polyurethane for the most part in medical devices and agricultural products where catalysts are unacceptable. A broader range of surfactants can be used in work with hydrophilic polyurethanes, and the surfactants determine cell structure. Table 3.4 shows the effects of surfactants.

We have sufficiently described the tools of polyurethane to provide a foundation. The choices of isocyanate and polyol dictate the compression and tensile characteristics. As we are about to show, the polyol serves another purpose. Through careful

TABLE 3.4
Control of Cell Size Using Surfactants

Surfactant	Cell/Foam Type
Pluronic L-62	Fine cell/wicking
Pluronic L-520	Hydrophobic/medium cells
Brij 72 (ICI Americas)	Ultrafine cells/super soft
Pluronic L-62 and P-75	Large cells/"sea sponge"

Source: Dow Chemical Company, 1978, Midland, MI, Hypol product literature.[26]

control of the reaction, the density and cell structure can also be used to serve the desired chemical purposes.

FACTORS AFFECTING CHEMICAL PROPERTIES OF POLYURETHANE

It is the primary intent of this book to teach readers how the chemistry of polyurethanes can be used to advantage. As we will show in a later chapter, the current polyurethane chemistries are effective for removing certain classes of organic molecules from the environment via a solvent extraction mechanism. The material presented above is useful for building an appropriate polymer system.

As we will also show, polymers built on hydrophobic polyols and isocyanates are appropriate for the extraction of hydrophobic pollutants. The intent of most polyurethane research is to develop polymers of sufficient strength to meet the requirements of a particular application, for example, designing a polymer to extract an aromatic hydrocarbon from the air. As we begin to develop applications, we will use the principles of solvent extraction and use specific polyols matched to extractants.

When a polymer system is intended to extract polar compounds, however, we deviate from conventional polyurethane research and focus on the chemistry of the material. In this area, standard texts on polyurethanes are of little value. The chemistry is extensive enough to require the rest of this book to describe. In keeping with the structure of this book, however, our discussion of achieving a desired chemistry will occupy only a few pages. Our intent is to show how chemically active aspects of a polyurethane are incorporated into a polymer. For the most part, these chemistries are hydrophilic. This reflects the work in our lab more than predetermined restrictions of the technology. We will illustrate why the tools discussed in this chapter are valuable. We will also discuss three categories of chemical modifications and how they are incorporated into polyurethane foam. The modifications are:

- Reservoir capacity
- Biocompatibility
- Ligand attachment

Reservoir capacity is the ability of a polymer to hold solutes within its matrix. Biocompatibility is the relationship of a polymer surface with biological materials, specifically living cells. Ligand attachment is the technique of attaching active side chains to the backbone of a polymer. The most interesting example is the covalent immobilization of enzymes without deactivation.

CONTROL OF RESERVOIR CAPACITY

We refer to a *reservoir* as the ability to absorb, adsorb, or entrap active ingredients into a matrix. We will differentiate this from simply wetting the surface of a poly-urethane by corona discharges or surfactants inasmuch as those techniques relate generally to all polymer systems. In the context of this book, reservoir capacity is analogous to a bottle filled with an active ingredient. In most cases, swelling is an indication of reservoir capacity. For example, a hydrophilic polyurethane will swell to roughly twice its volume after immersion in water. The swelling indicates that water penetrating the polymer surface resides within the matrix. Compare this to a conventional cellulose sponge that does not swell in water, indicating that the water is held on the inside surfaces of the sponge. Similarly, if a conventional hydrophobic polyurethane is immersed in acetone, it too swells, indicating the reservoir capacity of the polyurethane for acetone. Adjuncts to reservoir capacity are the tools designed to deliver active ingredients. We will describe such systems as containers with leaks.

Reservoir capacity is, in our view, an attempt by a polymer to dissolve. Because of cross-linking and molecular weight, the system does not fully dissociate into a true solution. Rather than dissolving in the normal sense, the polymer is said to swell in the solvent. Absorption of a solvent, water or organic, is a volumetric phenomenon controlled by the relative polarities of polymer and solvent. A nonpolar backbone is preferred for absorbing nonpolar solvents. The molecule we call polyurethane, however, is not entirely nonpolar but is close enough for use as an absorbing matrix.

When a solvent is polar, a reservoir that reflects that must be developed. Hydro-philic polyurethanes were specifically designed to serve as reservoirs for polar solvents, although the inventors did not express their ideas in that manner. In many commercial cases, reservoir capacity ("the bottle") is too large. The container swells as it absorbs. The nature of the material is to lose most of its physical strength as it swells. While we need the polarity, we also want to optimize the size of the bottle and strength of the material. The current library of hydrophilic prepolymers does not provide that flexibility, but we now know that we can build our own prepolymers with copolymers of propylene oxide (PO) and ethylene oxide (EO).

Much of the work related to environmental and medical devices required hydro-philicity. This property is inexorably bound to the swelling of the polymer in water. The amount of swell or hydrophilicity is controlled by the polyol used to build the polyurethane. In many cases, the design of the system requires a compromise of hydrophilicity and physical strength, and the choice of polyol is the chief tool. We stated earlier that block copolymers were suitable for intermediate levels of hydro-philicity. Table 3.5 lists a series of polymers and their equilibrium values. Each polymer is the result of proper selection of an EO or PO copolymer. The table can be used as a guide in designing polymers of intermediate hydrophilicity.

TABLE 3.5
Polymer Hydrophilicity

Class of Polymer	Equilibrium Moisture (%)	% EO in Block Copolymer with PO (1000 MW diol)
Hydrophobic	<6	0
Amicus hydrophilii	6–15	10
Mildly hydrophilic	15–60	30
Hydrophilic	>60	100

Other designations are possible. It has been said that anything above 20% should be considered a hydrogel if, when swollen by plasma, it does not initiate an inflammation response in blood.[27] Conventional hydrophilic polyurethanes such as those sold by LMI (St. Charles, MI) or Rynel (Boothbay, ME) have equilibrium moistures in the 60% to 70% range. Both examples use polyethylene glycol as a polyol. By increasing the molecular weight and closely controlling the cross-linking reaction, very high equilibrium moistures can be achieved. High molecular weight triols of ethylene glycols are convenient.

Again, the intent of this design process is to provide reservoir capacity for a polyurethane. While we have described a pseudosolution phenomenon, reservoir capacity may also involve the addition of solids. While the process is typically referred to as *entrapment*, it is nonetheless an important capability. It is, however, rarely necessary to change the chemistry to achieve function. Any of the formulations ranging from hydrophobic to hydrophilic cited in Table 3.5 would be suitable as an entrapping medium. In subsequent chapters, we will illustrate the techniques of entrapping activated charcoal, ion exchange resins, and even living cells. When a hydrophilic matrix is desired, this technique is particularly convenient because the ingredients can be mixed with the aqueous material and blended with the prepolymer.

While absorption is a valuable quality, more often than not absorption capacity is used to carry solutes into the matrix of a foam. If this is done after the foam is produced, the process is referred to as *imbibing*. It is sometimes advisable to add the solute to the water phase used to produce the polyurethane. As long as the solute is not reactive with the prepolymer, the solute is deeply imbedded into the foam matrix. The imbibing process has a tendency to place the solute closer to the surface. The purpose is usually delivery in a controlled manner. The simplest example is to imbibe a liquid soap into a foam that is then dried. The soap diffuses out when the foam is wet.

The polarity of the polyurethane can be used to extract components from the environment. Thus, reservoir capacity (more than simply the size of a device) is controlled by the polarity of the polyurethane and the proper choice of polyol.

BIOCOMPATIBILITY

The second characteristic of a polyurethane used as a specialty chemical is biocompatibility. We will discuss the techniques available to make a polymer system neutral with respect to living cells and methods of creating binding sites for attachment of certain types of cells including techniques that make polyurethanes hemocompatible.

One method of producing a biocompatible surface is to prevent adsorption of proteins. If proteins adsorb or otherwise become attached to a polymer surface, the attachment can interfere with the normal cell functions. The interaction of a polymer surface and blood is equally problematic. A component in blood known as the Hageman factor detects hydrophobic surfaces. The signaling involves attachment of the factor to the surface and by the process of attachment, the factor becomes activated. This is the first step in the inflammation response that can lead to rejection. Thus, the development of a hydrophilic surface with minimal protein adsorption may become a strategy for the development of compatible medical devices.

Braatz et al.[28] studied the adsorption of proteins on polyurethane surfaces. They coated silica and silicone tubing with a variety of chemistries. Although they found differences, the polyurethanes demonstrated minimal protein adsorption given sufficient coating thickness. Correlations between the relative number of hard (isocyanate) and soft (glycol) segments were drawn. More soft segments resulted in less adsorption.

Polyethylene glycols are well known as protein-compatible molecules when coated or grafted onto surfaces.[29] Both protein and platelet adsorptions to polyethylene glycol (PEG)-modified surfaces were shown to be reduced by PEG chains when attached to surfaces at one end of a molecule. Adsorption and platelet attachment were shown to be inversely proportional to the length of the PEG chain; 100 monomer units provided minimal adsorption and adherence.[30]

A study of protein resistance of terminally attached PEG chains was performed.[31,32] Steric repulsion between the PEG chain and the approaching protein, hydrophobic, and van der Waals forces were calculated. High surface density and long chain PEGs were found to favor protein resistance.

Amiji and Park[33] studied protein adsorption and platelet adhesion on surfaces coated with polyethylene oxide and polypropylene oxide block copolymers. Fibrinogen adsorption and platelet adhesion were studied on glass and polyethylene surfaces treated with PEG and the Pluronic series of block copolymers. Interestingly, more platelet adsorption was observed on the ethylene glycol-treated surfaces than with the more hydrophobic Pluronic series. Within the series, differences were noted as the hydrophobe-to-hydrophile ratio was changed. As the hydrophobe concentration increased (more PO relative to EO), the adsorption decreased. This appears contradictory in the light of other research, but could be explained as being caused by absorption of the protein into the hydrophilic portion of the molecule, providing sites for attachment. We know that we can control cell adhesion by imbibing fibronectin into a hydrophilic polyurethane. This can also be accomplished by copolymerizing the fibronectin into the hydrophilic polyurethane backbone. In either case, a low adsorption polymer can be made to appear to have adsorption properties. In our work to minimize platelet adsorption, the polymer is preswollen with plasma. It is clear from these references that one strategy to minimize protein adsorption is to use a polyethylene glycol as suggested by Braatz or use a member of the Pluronic family as suggested by Amiji.

We have already described how these polyols can be incorporated into a polymer by reaction with an isocyanate. It is clear that the reaction products of the isocyanate will affect the biocompatibility of the resultant polymer. Careful construction of a polymer with high-molecule-weight polyols offers some design control to minimize the effects of the isocyanate derivatives. We must also be aware that changing the polymer formulation also changes its process ability. Composites avoid this complication to some degree.

LIGAND ATTACHMENT

The last chemical characteristic is the attachment of ligands. Unlike the other properties (reservoir capacity and biocompatibility) that could be incorporated into a one-shot process, this aspect is most conveniently practiced at the prepolymer level. The philosophy is to use some of the isocyanate functionality of the prepolymer to attach active side chains. This is illustrated in two examples that we will discuss again.

Faudree[34] teaches the production of an alcohol-grafted polyurethane for the express purpose of absorbing waterborne oil spills. A C_{12} alcohol is reacted with a hydrophobic prepolymer. Care must be taken not to consume too many of the NCO groups because this will prevent the conversion of the prepolymer to a full polyurethane.

In a second example, Storey et al.[35] demonstrated that one could covalently immobilize amyloglucosidase using hydrophilic prepolymers. A 5-mg/ml solution of the enzyme was mixed with an equal volume of prepolymer. The method was judged superior as a support for enzyme immobilization. The percent activity immobilized in the polyurethane foams was $25 \pm 1.5\%$.

4 Extraction of Synthetic Chemicals

INTRODUCTION

In a 1999 report on the incidence of contaminants in public water systems,[32] the U.S. Environmental Protection Agency (EPA) stated, "Many past EPA studies have shown that some simple measures, such as population (or population density) are valid indicators of pollution, because it is human activity and land use — be it manufacturing or agriculture — that is the source of most pollutants, particularly the organic chemicals. Various demographic or other factors were evaluated as independent measures or indicators of pollution potential." In a report by the U.S. Geological Survey[1] on volatile organic compounds (VOCs) in groundwater, the authors went so far as to express the probability of finding contaminants (P) as a function of the population density (x) in units of persons per square kilometer.

$$P = \frac{e^{[-3.1+0.40\ln(x)]}}{1 + e^{[-3.1+0.40\ln(x)]}}$$

This relationship appears to be ominous and should serve as a charge to the technical community to address the relationship. It would be hopelessly optimistic to think we could break the relationship on an absolute basis, but we can certainly address the slope of the curve.

In the above equation, $0.40 \ln(x)$ represents the slope of the probability curve. Reducing it to $0.30 \ln(x)$ would translate to a 23% reduction in VOCs released to the environment.

The authors of both the EPA and Geological Survey studies used a convention that we will adopt. VOCs are compounds whose vapor pressures are high enough to be present in the air and water. Synthetic organic chemicals (SOCs) are organics that by their nature are significantly soluble in water. They are of particular concern inasmuch as their solubility permits dissolution in groundwater, thus increasing their mobility. Finally, inorganic chemicals (IOCs) include phosphates and chlorides whose impacts on the environment range from benign to corrosive. As we all know, certain inorganics can be toxic to plants and animals. Table 4.1 presents examples in each category.

The EPA regulates 64 chemical contaminants that are of primary concern based on their potential to cause chronic health problems. Some of the 64 are cited in Table 4.1. The EPA report[36] noted that all 64 contaminants were found in drinking water systems, but the frequency varied widely. Fifty-nine contaminants were reported

TABLE 4.1
Examples of Three Categories of Pollutants

Contaminant Category	Contaminant	Maximum Contaminant Level (mg/liter)
SOC	Alachor	0.002
	Atrazine	0.003
	Furadan	0.04
	Ethylene dibromide	0.00005
IOC	Antimony	0.006
	Arsenic	0.05
	Cyanide	0.20
VOC	Benzene	0.005
	1,2-Dichloroethene	0.10
	Toluene	1.00

Source: Adapted from Lesley-Grady, et al.[37]

at half their respective maximum contaminant levels (MCLs) and concentrations greater than MCLs were "not uncommon." Seven of the 21 VOCs on the EPA list (including ethyl benzene, trichloroethylene, and vinyl chloride) were found in every state examined.

Surface water was determined to be more vulnerable than groundwater for most contaminants. SOCs were more common in surface waters, and most of the contaminants that exceeded the MCLs were in surface waters. For VOCs, however, no significant difference was found in the number of contaminants exceeding MCLs in ground or surface waters.[1] IOCs were found to be equally common in both ground and surface waters. Many SOCs (pesticides, in particular) were seasonally present.

These studies dealt with the pollution of water resources. VOCs, as their name implies, contaminate the air as well. Odabasi et al.[38] reported on aromatic hydrocarbon pollution of the air in Chicago, and compared it with air pollution in Boston, London, and Houston.

The concentrations of VOCs in urban air and river waters were investigated by Rosell et al.[39] Samples were taken from the waters of the Besos and Llobregat rivers and from the air in Barcelona, Spain. The waterborne VOCs were extracted by air sparging, and the air samples were studied using charcoal and polyurethane foam adsorption extraction methods. In both types of samples, C1 to C5 alkylbenzenes and n-alkanes constituted the two major VOC groups. Chlorinated compounds were abundant in water samples. Tetrachloroethane was the predominant chlorinate in airborne VOCs. Differences in the efficiency of separation were noted as functions of molecular size. Polyurethanes were used for the high molecular weight samples while carbon was preferred for the smaller molecules. This difference has guided our work on the incorporation of carbon in a composite hydrophilic–hydrophobic polyurethane, thus making a broad polarity extraction system with a broad molecular weight response possible.

While problems with these contaminants are of constant and increasing concern, they are not the only environmental issues. Many of the contaminants mentioned above are not found in the average system. Most are of analytical interest, and therefore are not highly visible. This does not diminish their importance related to monitoring and remediation. Toxicity is pernicious in the sense that toxic levels are often below the limits of detection by normal human senses, and the effects of many contaminants on the list are chronic in nature.

Most of the contaminants discussed thus far are of industrial or agricultural significance. The development of our economy is based on equilibrium of supply and demand. The need for chemicals that improved agricultural efficiency led to the development of many of the SOCs on the list. The contribution of these chemicals is, for the most part, undeniable. It is obligatory for us to develop new technologies that can achieve some level of equilibrium between the need for efficient production of food and maintaining residual or fugitive levels of these SOCs below toxic (as opposed to unrealistic) concentrations.

Our consistent need to improve our daily lives also led to unanticipated industrial developments. For example, the production of automobiles led to expansion of the oil production (or vice versa) and metal working industries, both of which account for pollution by several compounds cited on the contaminant list. The chemical processing industry has been responsible for many items we now consider the essentials of modern life. From plastics to modern electronic devices, the chemical industry has guided and benefited from developments and also exerted colinear effects on the contamination of air and water. Again, the development of remediation technologies is needed to establish an acceptable equilibrium.

TREATMENT OF SANITARY WASTE

This chapter will explain how polyurethanes can play important roles in achieving acceptable equilibrium. Before we begin that discussion, we should discuss our most visible environmental problem — treatment of human waste. In the previous section, we covered the establishment of the equilibrium of SOCs, VOCs, and IOCs. The environmental effects of our daily existence resulting from personal hygiene and personal waste control are of immediate concern. Arguably, the types of pollution mentioned in the previous paragraphs warrant continuous monitoring. Human sanitary waste disposal requires continuous action.

Evidence of the need for continuous action can be found in the massive waste treatment plants in urban areas and septic systems in rural communities. In the developed world, the problem is under nominal control, and this allows us a complacency not evident in the rest of the world. In undeveloped countries, the return of water free of nutrients and bacteria to the environment is a goal to be anticipated and a luxury few can afford.

The need for safe water is illustrated by correlating the percentage of the population that has access to safe water and the mortality rates among children below 5 years of age.[40]

Around 1.5 billion people worldwide lack access to safe drinking water; about 80% of them live in rural areas and depend upon small or individual supplies. The

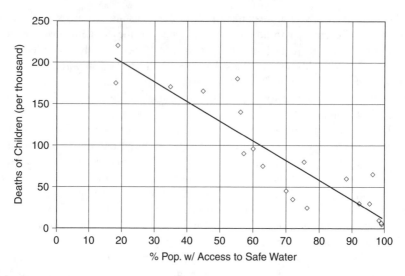

FIGURE 4.1 Child mortality by access to safe water.

effectiveness of interventions is variable and sometimes very low. Data from across the globe indicate that the effectiveness of small systems is measured in months rather than years. Because of the high rate of failure of these systems, projects focus on expanding systems into underserved areas and also maximizing the working lives of existing systems. Professional management is often unavailable to small communities and problems with operation and maintenance are major factors in the short working lives of the systems.

One of the issues is contamination by bacteria and viruses. Some of these organisms have long been recognized; others are newly emerging. Most often, this type of contamination results in diarrheal diseases that are among the top three causes of death worldwide and constitute the leading cause of death among children in developing countries.

The outbreak of cholera in Latin America that started in Peru in January 1991 was caused in large part by the lack of effective water treatment and disinfection. The result was 1.2 million cases and 12,000 deaths from 1991 to 1998.

The major concern of water utilities in the 1980s was dealing with coliform bacteria and the potential presence of Giardia cysts. By the 1990s, the list of pathogenic waterborne microorganisms had grown exponentially. Cryptosporidia, Microsporidia, Cyclospora, and Mycobacteria captured the headlines by causing water- and foodborne disease outbreaks.

The technology exists to solve the majority of these problems, but the solutions require skills unavailable in the Third World community. Building the capital systems common in developed countries is beyond the resources of Third World countries. Provision of effective technologies that use some of the techniques discussed in this book could be useful in this regard. Our laboratory, in part, is dedicated to this pursuit.

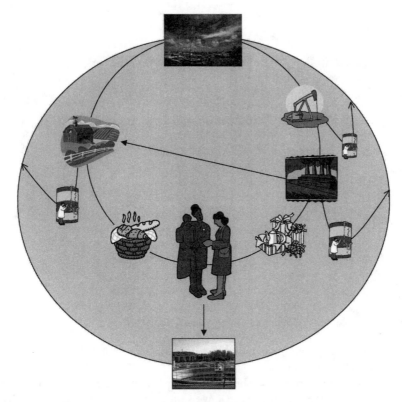

FIGURE 4.2 Closed-loop goal of environmental remediation and recycle.

The discussion of improved sanitary remediation systems is by no means limited to the Third World. In rural areas of the United States, bacterial contamination of water supplies occasionally makes headlines.

SECTION SUMMARY

We have described a world that strives for a constant state of equilibrium. We use our air and water resources to establish standards of living. In order to maintain those conditions, it is incumbent on us to ensure that what we extract from the environment is returned in a usable form. Inasmuch as we have no material communications beyond our planet, we are forced to work with the resources we have. We cannot destroy our water resources as we use them. We must understand that air is another finite commodity that must be returned and reused.

Many maintain that the equilibrium we have achieved mostly through ignorance, but occasionally through carelessness or even intentional breaking of laws, is unhealthy. In any case, few would object to the notion that seeking an equilibrium that includes safe water and breathable air is a desirable goal. While many would argue that technology has produced the unhealthy environment, it is not arguable

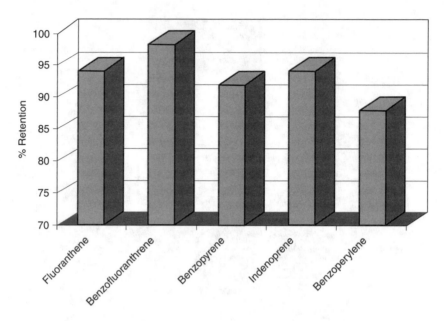

FIGURE 4.3 Extraction of airborne contaminants by polyurethanes.

that technology also represents the hope for a new balance between use and return of our water and air resources.

Figure 4.3 summarizes the objectives of environmental research. The air and water resources appear at the top. Conversion of resources via agricultural and industrial means is shown on the perimeter of the inner circle. The treatment of industrial, agricultural, and human waste in a closed system in order to return the components as air and water resources is shown in the outer circle.

The balance of this chapter focuses, for the most part, on the removal of contaminants from the environment by nonbiological methods. For engineering reasons, we will concentrate on removing VOCs and SOCs. The next chapter will discuss biological methods of remediation and water streams contaminated with VOCs, SOCs, and IOCs.

TREATMENT OF ENVIRONMENTAL PROBLEMS BY EXTRACTION

Environmental treatments for removing pollutants include *in situ* degradation with microorganisms and enzymes, use of biofilters, and extraction and sorption of the pollutants. These and other techniques will be covered in this chapter, but for various reasons, the extraction of contaminants is of particular interest primarily because extraction requires no particular pretreatment of the chemical. Air can be injected into the soil around an aquifer and recovered in sorption towers for concentration and removal from the environment.

Alternatively, water can be pumped from an aquifer through sorption columns and reinjected into the groundwater system. In this context, *sorption* means any process by which a fluid (air or water) is contacted with a material to which the pollutant has an affinity. The affinity can be a physical trapping modified by some form of surface energy or a solvent-extraction process based on enthalpic principles. The result is that the fluid is pumped through the sorption medium and the pollutant is reduced or, hopefully, eliminated from the fluid.

Despite limitations, the most common sorption medium is activated charcoal — a form of carbon treated in such a way as to open a large number of pores. The surface energy of the material and the pores combine to produce a material that can first attract and then trap small organic molecules. The attraction is via adsorption rather than absorption. Adsorption applies to attachment to the surface; absorption is a bulk effect. Extraction is a bulk phenomenon. Simply put, adsorption is a function of surface area while absorption is a mass effect.

Adsorb is an important word. When a material adsorbs something, it attaches by chemical attraction. The huge surface area of activated charcoal gives it countless bonding sites for the adsorption of chemicals. Thus, when certain chemicals pass next to the carbon surface, they attach to the surface and are trapped.

Activated charcoal is used to trap carbon-based impurities (organic chemicals) along with molecules such as chlorine. Inorganic chemicals (sodium, nitrates, etc.) are not attracted to carbon and pass right through. Therefore, activated charcoal filters will remove certain impurities and ignore others, but they stop working when all their pores are filled. At that point, a filter must be regenerated by reprocessing with steam. For some applications, regeneration is not possible and the material must be discarded. Additional problems include the fact sorption is based on molecular size. Pollutants whose molecules are bigger than the pores of the charcoal are unaffected. It has been reported that it was necessary to develop new carbons to extract methyl *tert*-butyl ether (MtBE), for instance. Regeneration, flow problems, and attrition of particles have been cited as difficulties. Activated charcoal columns are usually pressure vessels due to the large and dynamic pressure drops across the carbon bed.

Another technique is gaining interest because of the ease of regeneration and improved flow characteristics (small and constant pressure drops). Instead of physically trapping a pollutant in its pores, the technique involves direct attachment of the contaminant molecules to the sorption material, usually a polymer. All molecules are composed of a number of atoms with a confluence of electrons spinning around them in what is called an electrostatic field or electron cloud. The cloud, however, is not necessarily uniformly distributed.

Oxygen atoms are considered "electron withdrawing" while carbon atoms are "electron contributing." Thus, the electrons around a molecule containing oxygen and carbon will have higher electron density around the oxygen. Other electron withdrawing atoms include chlorine and fluorine. The molecules can be assigned values based on the nonuniformity of their electrostatic fields. The quantitative form of this differentiation is dipole moment; the qualitative characterizations are polar and nonpolar.

When two polar molecules (a solute and a solvent) approach each other, they can become attached (dipole–dipole interaction). If the solvent is water, the solute is said to be water soluble. Certain molecules, however, have little or no polarity. When they come into contact with a polar solvent (water, for instance), they cannot interact in any significant way and are said to be insoluble. Adding a nonpolar molecule to a nonpolar solvent (hexane, for example), produces an electrostatically neutral system and the materials mix uniformly. The energy associated with such a system is referred to as van der Waals forces. In any case, the effect is strong enough to be used commercially. In the real world, few molecules can be assigned a polar or nonpolar label; they have degrees of polarity.

All these types of solute–solvent associations are summed up in a rule of thumb learned by all chemists: like dissolves like. The chemical processing industry depends on the ability to separate a useful chemical from a solvent by an extraction process. If a chemist wants to extract nonpolar chemicals, he or she would use a nonpolar sorption material. The opposite is equally true. In a mixture of polar and nonpolar chemicals, the two classes of compounds could be separated from each other. The analytical techniques of gas and liquid chromatography are based on this principle. In applying this principle to an environmental issue, however, the fact that the pollutants have a range of polarities makes the system problematic. Thus, carbon is the material of choice because its affinity is based on molecular size, not on polarity.

We propose a system to extract pollutants from the environment based on the polarities of the molecules. In order to accomplish this, we must provide a solvent that is broad based with respect to its polarity. The solvent must also be of engineering quality in the sense that it must be able to withstand the stresses that will be applied to it with process scale equipment.

THEORY OF EXTRACTION

Extraction is a process for separating components in solution by their distribution between two immiscible phases. Such a process can also be called liquid extraction or solvent extraction. The former term may be confusing because it also applies to extraction by solid solvents. Since extraction involves the transfer of mass from one phase into a second immiscible phase, the process can be carried out in many ways. The simplest example involves the transfer of one component from a binary mixture into a second immiscible phase — extraction of an impurity from wastewater into an organic phase. In some cases, a chemical reaction can be used to enhance the transfer, e.g., the use of an aqueous caustic solution to remove phenolics from a hydrocarbon stream.

USES FOR EXTRACTION

In the chemical processing industry, extraction is used when distillation is impractical or too costly. Extraction may be more practical than distillation when the relative volatilities of two components are close. In other cases, the components to be separated may be heat sensitive like antibiotics or relatively nonvolatile like mineral salts. When unfortunate azeotropes form, distillation may be ineffective. Several examples of cost-effective liquid–liquid extraction processes include the recovery of acetic acid from water using ethyl ether or ethyl acetate and the recovery of phenolics from water with butyl acetate.

The most common form of extraction involves two immiscible liquids and a solute that is soluble in both and will be recovered. Liquid–liquid extraction requires (1) extraction, (2) medium recovery, and (3) raffinate desolventizing. The Udex process is a cost-effective liquid–liquid fractionation process for the separation of aromatics from aliphatics. The extraction solvent (diethylene or triethylene glycol) is recovered by steam distillation, and the raffinate and extract streams are desolventized by water extraction.

Process modifications use tetraethylene glycol as the extraction solvent and a mixture of light aliphatics and benzene as the wash solvent to the main extractor. Water condensate from the steam distillation is used to extract residual solvent from the raffinate and extract streams, so distillation for drying the extraction solvent is eliminated. Solids are removed from the recycled extraction solvent by filtration, while acids and high molecular weight fractions are removed by a solid adsorbent bed. Obviously, these processes are expensive and complicated. Most of the effort focuses on the recovery and not on the extraction.

MECHANISMS AND MATHEMATICS OF EXTRACTION

The principle of extraction is based on a physical chemical phenomenon known as partitioning. If two fluids, one of which contains a solute that is soluble in both, come in contact with one another, the solute will migrate from the original fluid into the other fluid. The extent (but not the rate) to which it will migrate is governed by the relative solubilities of the fluids. See Table 4.2. If the solute is equally soluble in both fluids, half will continue to migrate until the concentrations in both fluids are the same. If the solute is much more soluble in one fluid or the other, the fluid in which the solute is most soluble will accumulate most of the solute. For instance, if ethanol is dissolved in water and contacted with a solvent, the amount of ethanol removed from the water depends on the solvent.

The ratio is known as the partition coefficient and is a constant. For the most part, this ratio holds regardless of the concentration. The reason for this goes to the thermodynamic driving force to eliminate potential energy. It is an equilibrium constant and therefore obeys all applicable thermodynamic laws. One of these is a shift in equilibrium in response to temperature changes.

TABLE 4.2
Extraction of Ethanol from Water by Three Solvents

Solvent	Percent Ethanol in Water	Percent Ethanol in Solvent	Ratio
Butanol	25	75	3
Hexanol	50	50	1
Heptane	80	20	0.27

It is important to note that this discussion does not include the rate at which equilibrium is reached. Imagine that the two liquids are placed carefully without mixing. Equilibrium would eventually be reached but it would certainly take longer than if the two solvents were emulsified together.

One more aspect of extraction should be mentioned. Extraction obeys the thermodynamics of partitioning including the effects of temperature. The efficiency of extraction is dramatically affected by temperature. As the temperature increases, the relative amount of extractant in the phases changes. This is not the case with activated charcoal which operates by a trapping technique. From the perspective of regeneration, a thermodynamically controlled system would appear to be superior.

APPLICATION OF EXTRACTION PRINCIPLES TO REMOVAL OF ENVIRONMENTAL POLLUTANTS

The use of extraction as we have described it to remove pollutants from groundwater would be unworkable. The concentrations are so low that large quantities of water-insoluble solvents would have to be used and recovered. Inasmuch as the recovery of the solvents would be inexact, more pollutants would doubtless be released into the environment than would be removed. We may have found a solution to this puzzle by using a solvent that does not have to be recovered in the typical sense of the word.

It will be surprising to most engineers that basing most extraction systems on mixing two liquid solvents is not a requirement. *Perry's Chemical Engineer's Handbook*[41] contains many pages of extraction combinations, all of which are liquid–liquid systems. What we (and others) know, however, is that solid polyurethane can be one of the solvents. Nothing that limits the effects on liquids. Certain molecular considerations of solids (e.g., crystallinity, surface area) limit their use, but solids obey the same rules of extraction as liquids. Polyurethanes, however, have a unique advantage in that they are typically made into open-cell foams. This has the effect of increasing their surface area that is analogous to emulsifying two liquids for the purpose of extraction. An open-cell foam also combines high surface-to-volume ratios and high void volume — very important engineering characteristics. Certain forms of open-cell foams also have peculiarly low mass transport properties, specifically the ability to handle high flow rates with low resistance to flow. If these engineering descriptions are valid and an open-cell foam indeed acts as a solvent for extraction purposes, it would obey the thermodynamic principles of regeneration and therefore the problems of distillation and subsequent losses are avoided. It could therefore be considered an ideal solvent in this application.

As we will show in the next section, polyurethanes are capable of filling this role. The air pollution data presented in Table 4.1 were collected via extraction by polyurethane foam. The chemicals were released from the foam for analysis. While a problem related to the breadth of chemicals that can be extracted exists (conventional polyurethanes are essentially nonpolar), we will discuss how even this problem can be eliminated, thus making polyurethanes the materials of choice for groundwater and air remediation.

FIGURE 4.4 Improved absorption of oil by a derivitized polyurethane foam on right. On left, a reticulated pur.

EXTRACTION FROM AQUEOUS MEDIA

Gesser et al.[42] initiated the application of polyurethane foam for the extraction of organic contaminants from water. Since then, several published investigations have described the application of foams as extractors for chlorinated organic compounds, polycyclic aromatic hydrocarbons (PAHs), and other organics.

Saxena et al.[43] used polyurethane as a system to concentrate PAHs found in water. The samples were prepared by spiking water. After passing the water through the foam, the amount of PAH was measured. Polyurethane foam was used as an alternative to activated charcoal for the extraction of trace organic contaminants in water (Figure 4.4). Water was passed through the foam column at about 150 ml/min. It was noted that higher flow rates were used compared to the rates that could be used with carbon. According to the author of the study, the polyurethane foam was far superior to the active charcoal.

In another study, polyurethane was used to extract phenol. The extraction efficiencies of the foam were compared to those of other extractants including activated charcoal. The urethanes demonstrated the highest efficiencies (60% to 85%). Carbon was 45% effective.

Faudree's patent[34] teaches the production of an alcohol-grafted polyurethane for the express purpose of absorbing waterborne oil spills. By attaching long chain aliphatic hydrocarbons to a polyurethane prepolymer, they were able to increase the ability of the polyurethane to increase the absorption of 10W30 oil by about 20%. The patent, however, goes further by teaching that the polyurethane should be ground. We feel we improved the technology grafting Faudree's polyurethane to a reticulated polyurethane foam to produce a sheet of oil-absorbing foam. In our tests

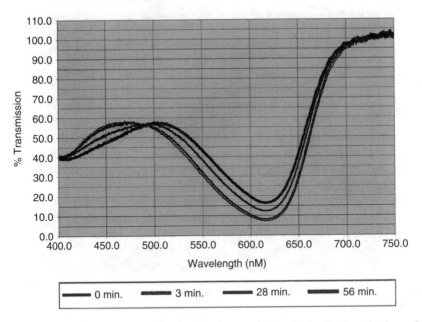

FIGURE 4.5 Extraction of Bromothymal Blue from water by a hydrophobic polyether polyurethane, indicating little or no extraction.

on 10W30 motor oil, we demonstrated its superiority over polyurethane. Figure 4.5 shows the visual comparison.

While the properties of polyurethane as an extractant are useful, several problems make it less than ideal. Polyurethanes are far more specific than activated charcoal in removing contaminants. Charcoal separates particles by size, and while it has some specificity, it is well suited for mixtures of diverse chemistries (PAHs vs. halogenated hydrocarbons). Polyurethanes, however, operate on the principle that "like dissolves like." They consist of hydrophobic isocyanates and hydrophobic polyalcohols. Thus, the molecules are hydrophobic. The polyalcohol backbone has some polarity, but it is hindered and therefore has a low net polarity. Inasmuch as the extraction effect is based, at least in part, on polarity, polyurethanes are most effective for nonpolar pollutants

EXTRACTION OF PESTICIDES

El Shahawi et al.[44] studied the use of polyurethane for the extraction of pesticides and similar compounds from water. In a 1993 paper, they reported investigations of the extraction of Dursban, Karphos and Dyfonate by activated charcoal and a polyether polyurethane foam. The first part of the study focused on isotherms. Weighed portions of carbon and polyurethane were placed in standard solutions of the pesticides and after a period of soaking and agitation, the concentrations of organics in the supernatant solution and in the extractant were measured. Table 4.3 summarizes their data.

TABLE 4.3

Concentrations of Various Pesticides in Equilibrium with Carbon and Polyurethane Foam

Concentration in Supernatant	Dursban		Karphos		Dyfonate	
	Carbon	Polyurethane	Carbon	Polyurethane	Carbon	Polyurethane
10	15	19	9	14	7	9
20	23	32	15	23	10	16
30	28	40	19	25	12	20

Although these data do not predict the removal efficiency of the extractant, they are reflective of the relative ability of the material to extract on a mass basis. Polyurethane foam was the superior extract in all three cases.

Most importantly, however, El Shahawi went on to hypothesize the reasons for differences in the degrees of extraction. The reasons included molecular weight, but El Shahawi also concluded that the polarity of the molecule was a contributing factor: "the smaller the dielectric constant of the absorbate, the larger the amount absorbed." Stated another way, as the polarity of a pesticide increases, the ability of a polyurethane to extract it decreases. El Shahawi used hydrophobic (nonpolar) polyether polyurethanes. If he had used hydrophilic polyurethanes, he might have concluded, as we now know, that "as the polarity of a pesticide increases, the polarity of the extracting polyurethane must increase." We also know that the broader the polarity range of the solutes, the more important it is to use a polyurethane with a broad polarity.

El Shahawi concluded that polyurethanes were preferred over carbon because of, among other properties, ease of regeneration. To support this, they determined the H values of a number of pesticides by determining partition coefficient as a function of temperature. The partitioning of a solute is in part defined by the vant' Hoff equation. The integrated form of the equation is:

$$P = \frac{\Delta H}{R}\left(\frac{1}{T}\right) + C$$

where P is the partition coefficient, H is the strength of the attraction of the solute and solvent, R is the universal gas constant, T is the temperature (in Kelvin), and C is the integration constant.

The slope of a plot of the partition coefficient vs. the reciprocal of the temperature (in Kelvin) is H/R. This is the fundamental equation of gas and liquid chromatography. In our laboratory, we coat a capillary column with a polyurethane of interest and measure the retention time of chemicals passing through it. The retention time is colinear with the partition coefficient.

To further explore the effect of the polarity of the molecule on its extractability, we studied the extraction of a water-soluble dye molecule by polyether polyurethane (hydrophobic) and a hydrophilic polyurethane (HPUR). The dye was bromothymol

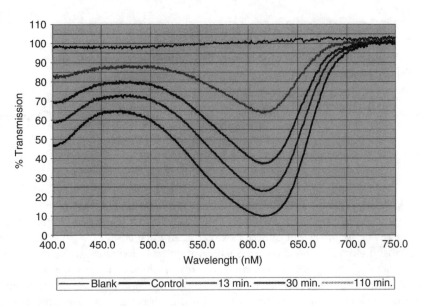

FIGURE 4.6 Extraction of BTB with HPUR.

blue (BTB), a pH-sensitive dye that changes from yellow to green to blue at pH levels of 6.0 to 6.5 to 7.0, respectively. We duplicated the experiments and measured the extraction effect by monitoring the intensity of the color in the supernatant with a visible spectrophotometer. If no extraction occurred, the percent transmission of the liquid would not change. If extraction from the solution occurred, however, the percent transmission would increase as a function of time as shown in Figure 4.5.

In a test of the ability of polyurethane to extract BTB (Figure 4.5), the absorption of light is unaffected by the presence of the polyurethane. Conversely, the absorption of light is strongly affected by the HPUR (Figure 4.6). This supports the hypothesis that as the polarity increases, the need for HPUR increases.

In another experiment using hydrophilic polyurethane, we showed the extraction of fabric dyes from washing machine water. Five extra-large men's sweaters (three dark blue, one red and one green) were washed using recommended procedures and a commercial detergent in a household washing machine. Several pieces of hydrophilic polyurethane foam were placed in the washer along with the sweaters. After a standard cycle, the hydrophilic foam was removed and the color measured by a visible spectrophotometer in the reflectance mode. Figure 4.7 depicts the results.

In an experiment that differentiates polyurethanes and HPURs, we measured the extraction of polar fragrance molecules from air. As seen in Figure 4.9, we placed coupons of HPUR and polyurethane foam on the pan of an electronic balance. Inside the chamber, we placed a pan containing a fragrance (Avon's "Fifth Avenue" essential oil, Shaw Mudge, Shelton, CT). The mass was monitored for 24 hours. The hydrophobic foam gained less than 1 mg of weight while the hydrophilic gained 42 mg. This experiment was conducted in support of the use of HPUR as a fragrance delivery system. An important aspect is the length of time that the system would release the fragrance. The odor in the sample persisted for over a month.

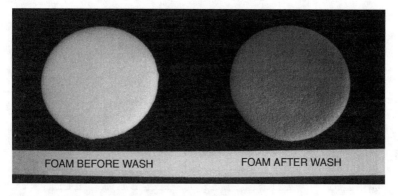

FOAM BEFORE WASH   FOAM AFTER WASH

FIGURE 4.7 Evidence of extraction of fabric dye by HPUR.

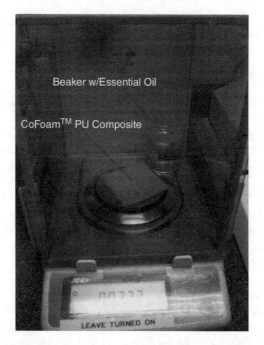

Beaker w/Essential Oil

CoFoam™ PU Composite

LEAVE TURNED ON

FIGURE 4.8 Polyurethane coupon in chamber of an analytical balance.

DEVELOPMENT OF BROAD-BASED EXTRACTION MEDIUM

It is, of course, possible to prepare a molecule that has both polar and nonpolar characteristics. This is the basis of surfactant chemistry. Typically, a nonpolar molecule is modified by sulfonation. The well-known Pluronic family of surfactants is based on block polymerization of polypropylene oxide (the hydrophobe) and polyethylene oxide (the hydrophile). It is conceptually possible to build a polyurethane

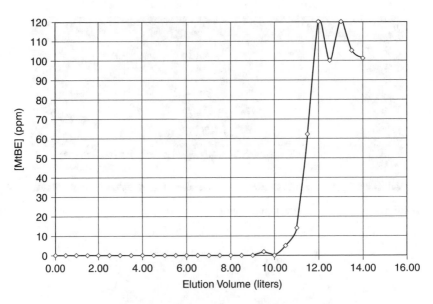

FIGURE 4.9 Breakthrough study of an MtBE solution through an HPUR column.

by combining hydrophilic and hydrophobic polyols, but doing so would diminish the structural and engineering characteristics of the product. Hydrophilix Corporation (Saco, ME), in an effort to build such a broad-range extraction solvent, developed and patented a composite that achieves the required range of polarity while still maintaining the engineering aspects; it is the reaction product of a hydrophobic isocyanate and a hydrophobic polyalcohol.

Thus, while it is possible to make a polyurethane that has a broad polarity structure, as we have often found while attempting to make polyurethanes of the correct chemistry, the physical requirements must be altered sufficiently to make the resulting device ineffective. In the case of a solvent extraction system for the typical range of groundwater pollutants, we would require broad polarity, excellent physical strength, high surface area, and high void volume along with the polarity of a hydrophilic polyurethane.

To address this paradox, we investigated and developed a number of composite materials. The most useful was produced by grafting a hydrophilic polyurethane onto the inside structure of a hydrophobic polyurethane. The process by which this is done is described in Chapter 3. We combined the polar nature of the hydrophilic polyurethane with the nonpolar nature of the conventional polyurethane.

CASE STUDIES

The nature of the problem dictates that studies on a real system are required to determine true efficacy. While we are involved in a field trial of CoFoam™ to remove methyl *tert*-butyl ether (MtBE) from groundwater, to date we have completed only laboratory and pilot plant testing. Nevertheless, the results point toward eventual success. (Registered trademarks of Hydrophilix Corp., Saco, ME, USA.)

All the data we will report here confirm the potential of the material as a broad-range extraction medium. In order to build interest, however, we will have to provide engineers with more information than our anecdotal studies can provide. The cited studies lend credibility to our hypothesis. As stated, only large field trials will ultimately confirm our hypothesis, but current studies provide sufficient reasons for optimism.

USE OF COFOAM TO EXTRACT MTBE FROM WATER

We sponsored a study by the environmental laboratory of the University of Maine in Orono. The study determined that HPUR had the capacity to extract about 3.5 mg of MtBE per gram of hydrophilic foam. The university and others attributed this to a solvent extraction mechanism based on the fact that the extraction units are noted in milligrams per gram of foam. If the reaction were purely a surface phenomenon, the units would have been stated in milligrams per unit of area.

We were encouraged by the results and therefore expanded the effort to include continuous flow of an MtBE solution through columns of hydrophilic foam. CoFoam has the strength and flow-through properties that permit its use in an engineering environment while it retains the functionality and unique properties of an HPUR. In this application, high flow rates with low pressure drop were the properties of primary interest. Accordingly, the pilot studies were done using columns of CoFoam meeting the following parameters:

- Length of column = 280 cm
- Inside diameter = 4.0 cm
- Volume = 3.5 liters
- Foam capacity = 420 g (typical)
- Weight of hydrophilic foam = 300 g (typical)
- Flow rate = 0.05 l/min (typical)
- Temperature = 15 to 45°C (typical)
- Concentration = 100 ppm MtBE
- Equipment = fixed-rate peristaltic pump; control and data acquisition via workbench on Macintosh platform; gas chromatograph and UV–VIS spectrophotometer for analysis

The procedure involved pumping an MtBE solution through the columns and analyzing the effluent to detect the first MtBE elution. This was termed breakthrough and was used to calculate the capacity of the CoFoam. Figure 4.9 shows the results of one determination.

From these data, we determined that the column could extract about 3.5 mg MtBE per gram of HP, confirming the findings of the University of Maine. We drained the columns and purged them overnight with air heated to 65°C. The MtBE solution was again pumped through the columns and the breakthrough determined. We found that the columns recovered the capacity to adsorb ca. 3.5 mg/g. The ability to regenerate the column was thus confirmed. The mild regeneration treatment would appear to present an advantage over activated charcoal that requires more complex processing conditions.

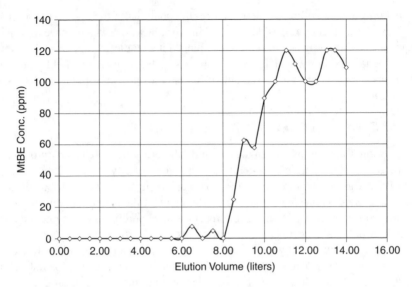

FIGURE 4.10 Breakthrough at 35°C.

The column was again purged with warm air to regenerate. MtBE was pumped through the column but at a higher temperature. It was thought that the process was in equilibrium and therefore subject to the normal characteristic of a partition coefficient phenomenon. The breakthrough was determined with the MtBE solution at 35°C. Figure 4.10 presents the data. In this final experiment, the capacity of the foam was reduced to 2.8 mg/g of HP foam.

COMBINATION OF CARBON ADSORPTION AND ENTHALPIC EXTRACTION BY POLYURETHANE

We attempted to develop a multipurpose extraction system by combining the carbon adsorption and polyurethane extraction techniques. Our research continues. Our first experiments focused on the production of a carbon-impregnated reticulated foam. Several companies promote this technology, but in our tests, the method of incorporation of the carbon significantly reduced the activity of the carbon.

An activated carbon (Nuchar RGC Powder, 879-R-02) was supplied by the Chemical Division of Westvaco, Covington, VA. A slurry of the carbon was made in water and emulsified with a TDI-based hydrophilic polyurethane prepolymer (Hypol 2002, Dow Chemical, Midland, MI). Immediately after mixing, the emulsion was grafted to a 30-ppi polyether polyurethane (Crest Foam T-30, Monachie, NJ.). The amount of carbon was determined gravimetrically to be 29% by weight.

A 1-liter polyethylene bottle was used for the extraction studies. A small amount of butane (approximately 0.2 g) of butane was bled into the bottle. The bottle was capped and a 2-ml sample was withdrawn by inserting a syringe needle through the side. The sample was then injected into a gas chromatograph. The timer was started and a chromatogram recorded. Samples were taken for 2 h to ensure that the butane was not affected by the bottle. Figure 4.11 depicts the result.

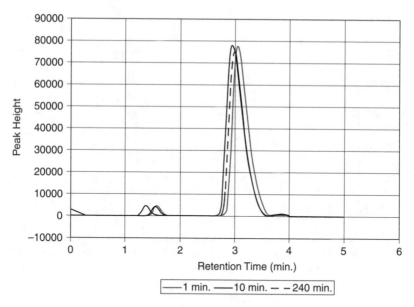

FIGURE 4.11 Concentration of butane in a polyethylene bottle.

A study sample was introduced into the bottle, capped, shaken for 10 sec, and a sample of the gas taken and analyzed as above. A decrease in butane peak height was considered evidence of extraction. A plot of the butane peak height vs. time from introduction of the sample (reaction time) was considered a measure of the kinetics of extraction. The data were fitted to an equation of the form:

$$C_{\text{butane}} = A(R_T)^{-B}$$

where C_{butane} = butane peak height, A = calculated intercept (function of initial concentration), R_T = residence time of carbon in the bottle, and B = slope of the line (rate of extraction). Activated carbon was added to the bottle and the butane concentration was monitored. Figure 4.12 is the chromatogram. Figure 4.13 plots peak height vs. reaction time based on the equation of the curve fit calculations.

The bottle was charged with butane again and a sample of the ungrafted reticulated foam was placed in the bottle. Figure 4.14 shows the kinetics of extraction. The extraction ability of the foam is significantly lower than the ability of the carbon.

A similar result was seen in the performance of a sample of hydrophilic polyurethane-grafted reticulated foam without carbon. (Figure 4.15). Figure 4.16 shows the analysis of the effect of the carbon-impregnated foam. It is clear from these data that the carbon-impregnated foam showed significant improvement in effectiveness. In all the extractions cited below, the CoFoam contained about 1.7 g carbon.

In an effort to confirm this extraction effect, samples of CoFoam were produced with increasing amounts of carbon. The amount of carbon was not determined for each sample, but we were able to show the relative amounts by determining reflectivity to visible light. The kinetics of extraction are shown in Figure 4.17.

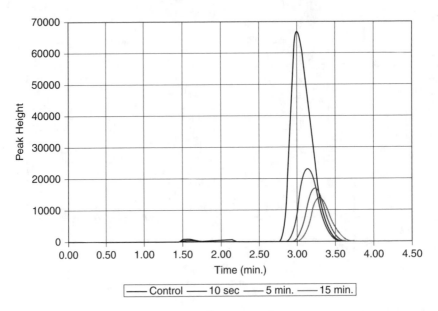

FIGURE 4.12 Extraction of the butane by Westvaco carbon.

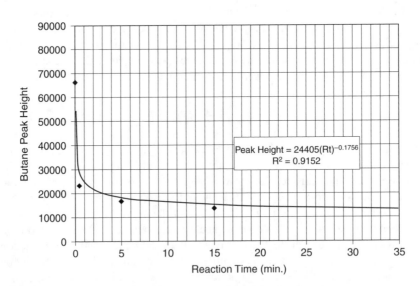

FIGURE 4.13 Kinetics of the extraction of butane by Westvaco carbon.

The slopes of the curves show an exponential rise consistent with the increasing volume percent of carbon that would increase the probability of finding carbon at the surface of the CoFoam (Figure 4.18). Plotting the slope against the reflectance data (which is related to the carbon concentration) yields results shown in Figure 4.19. Table 4.4 summarizes the data.

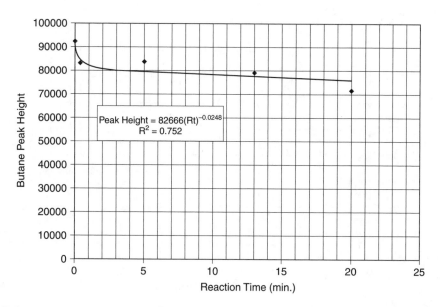

FIGURE 4.14 Kinetics of extraction of butane by a reticulated polyurethane foam.

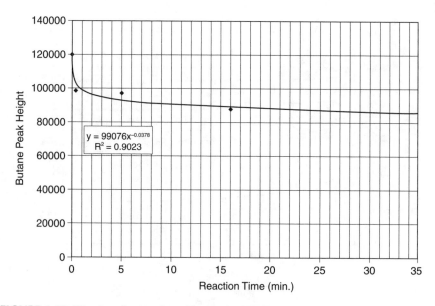

FIGURE 4.15 Kinetics of extraction of butane by HPUR.

Research continues on this technology. It is hoped that a carbon impregnated cofoam will eventually demonstrate the required carbon adsorption and solvent extraction effects.

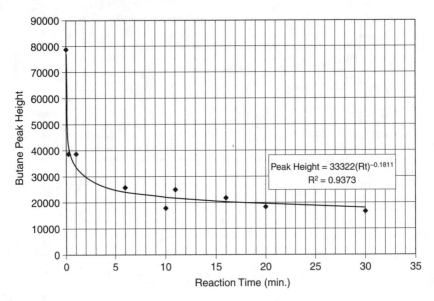

FIGURE 4.16 Kinetics of extraction of butane by HPUR impregnated with Westvaco carbon.

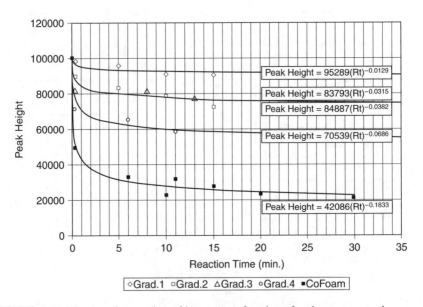

FIGURE 4.17 Kinetics of extraction of butane as a function of carbon concentration.

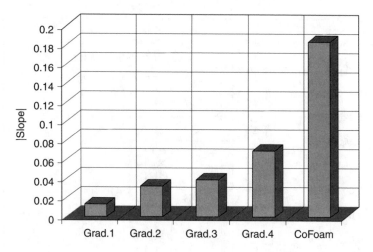

FIGURE 4.18 Slope of the extraction kinetics by carbon concentration.

TABLE 4.4
Slopes of Extraction Kinetics Curves

Sample	Slope of Kinetics
Westvaco carbon	0.176
Reticulated foam	0.025
HPUR-grafted reticulate	0.038
CoFoam (29% carbon, 0% reflectance)	0.181
CoFoam (5% reflectance)	0.069
CoFoam (12% reflectance)	0.032
CoFoam (19% reflectance)	0.038
CoFoam (30% reflectance)	0.130

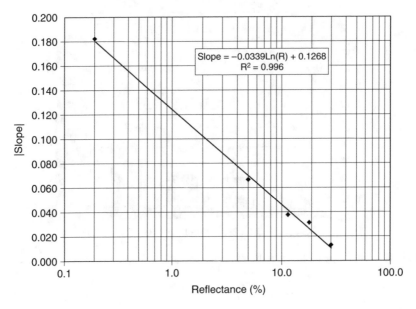

FIGURE 4.19 Slope of kinetics of extraction by carbon concentration.

5 Additional Environmental Applications

In Chapter 3, we introduced the three features on which the specialty chemical nature of polyurethanes depends:

- Reservoir capacity
- Ability to be colonized (biocompatibility)
- Ligand attachment

We described reservoir capacity as the ability to contain an active ingredient within a matrix. In this sense, we give the term *active ingredient* the broadest possible meaning. We will show how polyurethanes are used to absorb exudates from deep tissue wounds. The exudates are considered active ingredients. We likened reservoir capacity to a bottle and controlled release to a bottle with a leak. A polyurethane can serve as a controlled release device, and we will illustrate this in a number of applications.

A special class of reservoir capacity known as extraction was introduced in Chapter 4. Thermodynamically, the process is no different from the features described above, but from a use point of view, the process represents a special kind of bottle. Instead of leakage via diffusion caused by a potential energy difference, the leak arises from a shift in the partition coefficient from a change in temperature or in the solvent environment.

While our focus for this chapter will be on the colonization and the attachment of ligands, we will never entirely leave the topic of reservoir capacity. We discussed earlier the use of polyurethane as a scaffold for the development of colonies of microorganisms. We will use the capacity of polyurethane to hold solutes, both as a reservoir for nutrients and as a buffer, moderating swings in biological load, for example.

In addition to continued emphasis on reservoir capacity, we will shift our attention to other important specialty characteristics of polyurethane: the abilities to be colonized by living cells and to attach active ligands. More specifically, this chapter will illustrate how polyurethanes are used to address environmental problems. The problems we will discuss are different in nature from the problems discussed in the last chapter. This chapter focuses on the treatment of waterborne human waste.

The purpose of wastewater treatment is to remove pollutants that can negatively affect the aquatic environment before it reenters the natural milieu. The natural degradation of organic waste, whether in a dedicated facility or in a natural environment, is accomplished by biologically oxidizing pollutants. The process removes molecular oxygen from the water, thereby destroying its ability to support higher life forms such as fish. Most pollutants are carbon based and are said to be oxygen demanding. Nitrogen-containing pollutants are also of environmental interest and modern wastewater treatment facilities oxidize ammonia to nitrates. As we will show, both these oxidation processes are performed by microorganisms.

As population and industrialization continue to expand, eutrophication has become a problem due to the accelerated aging of bodies of water by the excessive growth of plants and algae attributed to discharges of nitrates and phosphates. These pollutants have imposed demands on engineers to develop cost-effective systems that can eliminate them from water.

In addition to the toxic chemicals that enter wastewater, treatment facility operators must be aware of other toxic substances. While most are handled by resident microorganisms, many are said to be recalcitrant and cannot be treated efficiently without special considerations. We will discuss the biodegradabilities of certain chemicals with regard to the special treatment protocols that may be required.

Systems that can treat these carbon- and nitrogen-based pollutants are well developed in the U.S., but are not common in the so-called Third World. Waterborne microorganisms represent the leading causes of mortality among children under 5 years of age. Although such toxic organisms are conveniently degraded by modern water treatment protocols, development of appropriate technologies to communities in need requires a different set of technological standards and a redefinition of cost-effective measures.

It is useful to begin our discussion of the treatment of ordinary wastewater by explaining how microorganisms are used to remove carbon and convert nitrogen-based pollutants. While arguably the most important part of the process is the direct action of organisms on pollutants, it is important to take a more system-oriented view. The system is composed of many physical and chemical phases and involves soluble and insoluble organics and inorganics. The phases differ in density. Some are hard; others are soft and hydrogel-like. Some are emulsified by natural surfactants. Some microorganisms residing in water are smooth and compact and settle easily while others are "fluffy" and have cilia that resist settling. The system that addresses these multiple pollutant forms must be rugged, durable, and versatile. Each process is handled by specific unit operations. Figure 5.1 is a typical flow diagram of a modern wastewater treatment facility.

Inasmuch as the system depends on the biological oxidation of matter and because the action on solid material is very slow, every attempt is made to remove solids from the waste stream. Settling basins provide a convenient method to accomplish the separation of higher-density, typically inorganic materials (sludge). After separation, the sludge is removed and treated as a separate waste stream. It can be mixed with other sludge-type wastes and composted — a biochemical process in its own right.

The remaining organic matter is fully saturated with water and therefore of neutral density or becomes water-soluble and processed by biochemical means. What is not degraded is often entrapped within the biomass and removed during the final

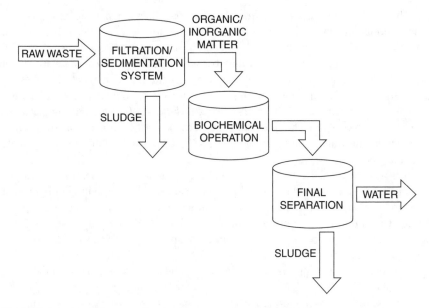

FIGURE 5.1 Flow diagram of typical wastewater treatment facility.

separation. The supernatant of the final separation is relatively clean and may be discharged without further treatment.

The success of a remediation process for the treatment of wastewater is a combination of three aspects. Each of these is instructive as we move toward a description of how polyurethanes can contribute to this technology. For the most part, municipal systems that remove pollutants from water are large successful chemical processing units. It is not our goal to challenge such operations. We intend to focus on highly efficient small systems that treat, for example, industrial and agricultural waste and can handle some of the demand on the municipal systems. In many areas, organic loads in waste streams are assigned to industrial dischargers and surcharged based on the amount of oxidizable carbon. It may be economically viable for these dischargers to set up highly efficient pretreatment of waste streams to remove some of the carbon.

BIOCHEMICAL CONVERSION

Soluble organic matter becomes food for microorganisms. In a typical municipal waste treatment system, colonies of organisms degrade organic waste by consuming it and converting it to CO_2 and biomass. The CO_2 evolves as a gas and the biomass is removed by sedimentation. The nature of the biomass is important. Biomasses are proteinaceous materials and fully hydrated. In many cases, they take the form of a slime rather than discrete organisms. Nevertheless, depending on the organism, the biomass separates into a packed layer or a fluff. In both cases, insoluble organics that passed the original physical separation are trapped within or by the biomass and are therefore settled out as well.

The two general classes of microbials are aerobic bacteria that depend on oxygen to treat pollutants and anaerobic bacteria that operate without oxygen. Most of the activities described above are aerobic. The conversion of organic material to CO_2 requires the fixation of oxygen. In the absence of dedicated wastewater treatment facilities, this need for oxygen would deplete the natural environment, thus making lakes and ponds uninhabitable by higher life forms. Aerobic transformations are typically used when organic loads are between 50 and 4000 mg/l. At higher concentrations, anaerobic techniques are used because of the difficulty in sufficiently oxygenating the water. The anaerobic process has the advantage of producing usable methane gas.

It is also necessary to treat noncarbon-based pollutants. We mentioned nitrogen and phosphorus as possible targets for biological remediation. Nitrogen is present in wastewater streams as ammonia. Conventional methods convert the ammonia to nitrate. Under normal circumstances, the conversion is sufficient. However, excessive amounts of nitrates contribute to the eutrophication of lakes and ponds. In these cases, a second class of organism is used to convert the nitrates to nitrogen gas.

Phosphates are also problematic. Most phosphates are present in residential water as inorganic phosphates, but organic moieties are common. Biological action converts the various forms into orthophosphates that can be removed from water streams by incorporation into specialized bacteria and algae.

BIOCHEMICAL REACTORS

The biochemical reactions cited above and the preliminary steps of removing heavy sediment comprise unit operations in a larger system. The designs, methods, and materials used in such operations define the quality of the results. The dewatering and sedimentation operations are discussed in engineering texts. Our interest is in the two general types of vessels in which biological conversion takes place.

In the first type, the microorganisms are suspended and therefore flow freely through the water. In the second method, the organisms are attached to a solid support. When a suspension method is used, the organisms are continuously stirred to prevent settling (until separation of the biomass is needed). In contrast, the attached organisms form a film on a solid support, and the fluid passing over it is acted upon by the immobilized cells.

SUSPENDED GROWTH BIOREACTORS

The simplest suspended growth bioreactor (SGB), in concept at least, is the stirred reactor in which polluted water is continuously added to a vessel colonized with microorganisms. Various overflow and underflow streams are mixed and recycled to create a continuous system of organic degradation and separation. The biomass ultimately is removed as sludge and is composted to permit its use as fertilizer, thus partially closing the environmental circle. Connecting several SGBs in a series provides additional flexibility. In times of heavy demand, additional SGBs can be brought on line.

ATTACHED GROWTH BIOREACTORS

As the name implies, the organisms in attached growth bioreactors (AGBs) are attached or immobilized on a support structure. Several types are used in commercial operations including packed columns, rotating discs, and fluidized beds. A polyurethane represents an important new category: a scaffold-immobilized bioreactor (SIB). In conventional reactors of the current type, the microorganisms typically reside on a plastic in a packed column arrangement. If the column is to react aerobically, the polluted water flows downward through the column. The oxidation is high at the top of the column where the carbon-containing pollutants are highest in concentration. This results in excessive and uneven growth of the biomass that can hinder flow. In practice, this is mitigated by recycling back into the inflow (diluting it) and adjusting the flow rate through the column.

The buildup of biomass is always a concern, however, and various techniques are used, typically based on sloughing off the organisms using the shear force of the liquid flowing through it. If the biomass is not removed from the recycle before adding it into the inflow, the reactor may be considered a continuous-flow type based on kinetics. If the biomass is removed, the kinetics can be described by the population of microorganisms immobilized on the support. In any case, the growth and removal of biomass are critical factors in AGB reactors. When AGBs are used in anaerobic mode, the polluted water flows upward or downward and the column is completely submerged.

Two other techniques use attached microorganisms for remediation. In a fluidized-bed method, the microorganisms are attached to particles dispersed throughout a continuously stirred reactor. In another arrangement, the microorganisms are attached to rotating discs.

Table 5.1 summarizes the general categories of processes and the common names that describe particular engineering designs.

TABLE 5.1
Comparison of Two Remediation Methods

Suspended Growth Bioreactor (SGB)	Attached Growth Bioreactor (AGB)
Activated sludge	Fluidized bed
Biological nutrient removal	Rotating biological contactor
Aerobic digestion	Trickling filter
Anaerobic Digestion	Packed bed
Upflow anaerobic sludge blanket	Anaerobic filter
Anaerobic contact	
Lagoon	

Source: From Lesley-Grady, C.P., Daigger, G.T., and Lim, H.C.[37]

This section represents a brief overview of the technology for returning polluted water to the environment. To some degree, the thousands of municipal wastewater treatment facilities in the developed world are unique in that they address local problems of flow, geology, and population density.

BIOCHEMICAL PROCESSES

Regardless of the engineering environment in which microorganisms exist, consumption and transformation of the pollutants by the organisms serve as the engines of the system. The intent of the engineering is to take full advantage of the abilities of the microorganisms.

From a remediation perspective, three types of organisms are important: Bacteria, archaea, and Eukarya,[45] archaea and bacteria are microscopic and lack nuclei. Eukarya organisms have nuclei. The primary actors on carbon-based pollutants are Bacteria, but Eukarya also have roles.

By various processes, including the production of extracellular enzymes, bacteria in the presence of oxygen consume and transform carbon-based pollutants. The transformation consumes oxygen and converts the carbon content of the pollutant to carbon dioxide. Part of the carbon is also used for cell synthesis. These are the dominant activities in a biochemical reactor.

Bacteria are also responsible for the degradation of inorganic pollutants, specifically ammonia and nitrites. The so-called nitrogen cycle begins with the biochemical conversion of ammonia into organic nitrogen compounds. This operation is called nitrification. The organic nitrogen compounds are further converted biochemically into nitrates. Since high nitrate levels contribute to algae growth, further treatment is required to convert them to nitrites and nitrogen. This is achieved in an anaerobic environment by the actions of denitrifying bacteria. Pseudomonas, an organism we will discuss later in this chapter, is a common nitrifier. Not all bacteria are beneficial, however. Some forms have filaments or cilia that inhibit settling, an important aspect in biomass removal.

Archaea are peculiar in that they are capable of growing in extreme environments. They are found at temperatures as high as 90°C and in solutions of high ionic strengths. One remarkable aspect of Archaea is that they also function in normal biochemical environments. They are currently used in anaerobic systems because of their ability to produce methane, but their applications are much broader. They also function as cometabolism systems by partially degrading hydrocarbons that are then acted upon by more common bacteria. Eukarya operate in systems where the pH is low and/or the oxygen level is deficient. Figure 5.2 summarizes the process with an emphasis on the evolution of gases and the production of biomass.

The evolution of noxious gases from a wastewater treatment facility is an unfortunate result of these processes. A model that shows the complete conversion of organic waste to CO_2, biomass, and water is idealistic. In practice, a number of volatile organic and inorganic gases evolve along with the CO_2. The dominant gas in the process is hydrogen sulfide (H_2S). We will discuss its conversion to sulfates, thus removing the odor-causing component from water. This is most conveniently done in a biochemical reactor (bioreactor).

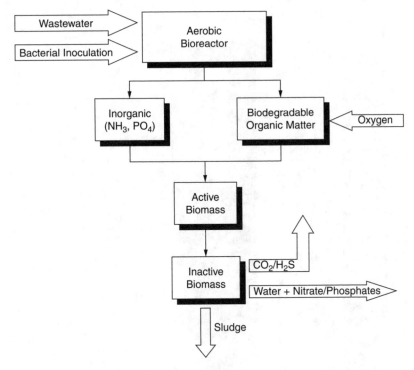

FIGURE 5.2 Aerobic digestion of organic waste.

DEVELOPMENT OF BIOFILM IN ATTACHED GROWTH BIOREACTOR

The microorganisms present in an attached growth bioreactor form a film on a solid surface rather than being suspended in water as occurs with suspended growth bioreactors. Once fixed to the surface, the wastewater flows over the surface and is degraded by the digestive actions of the organisms.

The association of the microorganisms and the surface involves the development of a film of hydrogel that entraps the organisms. The development of the biofilm (commonly referred to as a slime) is the process by which the organisms attach themselves to a surface. It is important to the remediation process to define the nature of the slime because the slime creates a moderating layer through which the carbon-based pollutants must diffuse for the degradation process to proceed. Much research relates to embedding the organisms in synthetic gels including polyurethane, and diffusion through these artificial attachment media must be considered.

The film that forms on the surface is very complex physically and microbiologically, but it is useful to conceptualize the film as a simple model (see Figure 5.3). The cells are embedded in this hydrogel coating. The substratum (left) will be defined later. For now we will consider it inert and nonabsorptive. The biofilm layer is a complex composite of the microorganism and a hydrogel vehicle. The phase immediately in contact with the biofilm is called the boundary layer.

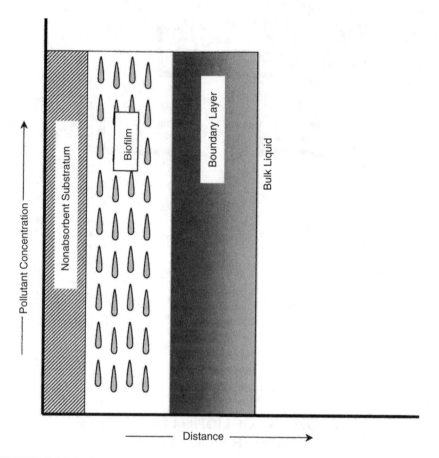

FIGURE 5.3 Idealized biofilm model.

Flow dynamics predict that flow through a pipe is nonuniform with regard to velocity across the diameter of a pipe, for instance. The flow at pipe walls is assumed to be zero. In our idealized biochemical reactor, this concept is represented by a boundary layer in contact with the biofilm. It does not have, of course, a discrete dimension. Rather, it is represented as an area in the structure that has reduced flow and therefore different kinetics than what we would assume exist in a bulk liquid. The boundary layer is affected by turbulence and temperature and this is unavoidable to a degree. Diffusion within the boundary layers is controlled by the chemical potential difference based on concentration. Thus the rate of transfer of pollutant to the organisms is controlled by at least two physical chemical principles, and these principles differentiate an attached growth bioreactor from a suspended growth bioreactor.

As a polluted fluid enters the environment of an attached growth bioreactor, a concentration gradient develops. For the purposes of simplicity, assume that we are passing water with a single contaminant through a bioreactor with a single degrading organism populating the biofilm. The carbon-based pollutants in contact with the

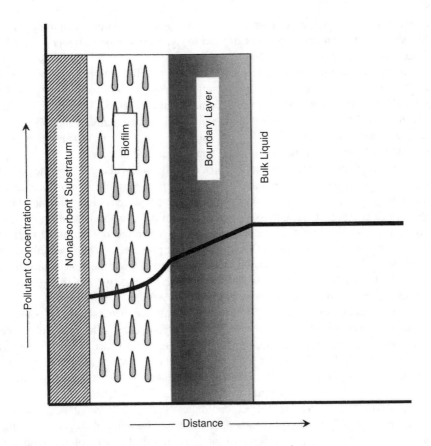

FIGURE 5.4 Concentration of pollutant vs. distance from substratum.

biofilm diffuse inward and are degraded by the organism. This decreases the concentration in the biofilm according to depth and reduces the concentration in the boundary layer accordingly. Figure 5.4 depicts this action and shows the effect of depths throughout the boundary layer and biofilm.

The thickness of the boundary layer is in part a function of the turbulence. The extremes range from the thickest in laminar flow to the thinnest in turbulent flow. The goal of the engineering of a flow-through attached growth bioreactor is to develop as much turbulence as possible without disrupting the biofilm by shear forces. As the turbulence increases, the concentration the biofilm "sees" increases, thus improving the efficiency of the system.

It is now useful to examine factors that affect the rate of diffusion, or more correctly, the amount of material that passes through the boundary layer and the biofilm. The mathematical principle that controls the rate at which the solute passes through a membrane by diffusion is Fick's law:

$$\frac{dm}{q} = D \frac{dc}{dx} dt$$

where dm/q = the quantity of matter passing through the biofilm per unit area, D = diffusion constant in grams per unit of time for a concentration difference of unity, dc/dx = concentration gradient, and dt = time.

In practice, the diffusion constant is modified to reflect the complex nature of the biofilm. In the development of synthetic hydrogels, the hydrophilicity of the polymer in part defines the concept of equilibrium moisture. We discussed this concept earlier when we described determination of equilibrium moisture. The first practical application is the diffusion constant. As the equilibrium moisture approaches 100%, the diffusion constant approaches that of water.

If polyurethanes are used to entrap cells, the diffusion will depend on the polyol used to build the polyurethane since the polyol defines equilibrium moisture. Later in this chapter, we will discuss a number of entrapment systems, including acrylates and polysaccharides. Each has its own equilibrium moisture and therefore unique diffusion constant. Only polyurethanes, however, offer the opportunity to affect changes in the constants. Conventional hydrophilic polyurethanes have equilibrium moisture levels around 70%. It is possible, however, to increase the molecular weight of a polyol (an ethylene glycol of 1000 molecular weight) to 3000 or more. This increases the equilibrium moisture to greater than 90%.

In the selection of a hydrogel, however, an additional factor should be covered. Hydrogel membranes are said to have a molecular weight cutoff. That is, the diffusion of higher molecular weight species (biopolymers, for instance). The molecules become bound by the polymer of which the hydrogel is composed. As the equilibrium moisture increases, fewer polymer chains are present to obstruct the flow of solutes through the gel. Higher equilibrium moistures have high molecular weight cutoffs.

It is conceivable that this effect could be used to differentiate degradation processes by molecular weight. An upper limit can be reached, however, as the equilibrium moisture approaches 100% and the tensile strength of the hydrogel approaches zero. As an aside, hydrogels are three-dimensional structures and therefore are cross-linked polymers. Cross-linking affects equilibrium moisture negatively. This leads to the correct conclusion that the highest diffusivity is achieved with weak gels. The goal of the polymer designer is to make the material strong enough to resist the shear forces that act on it. Our work focuses on the adhesion of the gel to the substratum and the cohesive forces within the gel.

The degradation of a pollutant can be accelerated by increasing the useful area of the attached growth bioreactor. Flux rate is the mass per unit of area per unit of time. Thus, as area increases, the mass per unit of time increases. Time can influence choices indirectly. In a packed column, one of the defining design features is called the empty bed residence time (EBRT), defined as the volume of the vessel minus the volume taken up by the substratum plus the biofilm.

We mentioned void volume earlier and it now becomes important. If the substratum has a very high void volume, the size of the tank in which it exists can be smaller than it would be if the void volume were lower. Thus, in a given space, a larger EBRT yields a higher flow rate for a given flux or a longer residence time. Both factors improve the rates at which pollutants are degraded.

With regard to the area, we should consider the materials used for column packing. Table 5.2 lists some common materials.

TABLE 5.2
Column Packing Materials

Type of Material	Surface Area (M²/M³)	Void Volume (%)
0.5" Rock	420	50
0.5" Ceramic	364	63
0.5" Carbon	374	74
0.5" Berl	466	63
1.0" Raschig ring	190	73
1.0" BioBall	160	—
Reticulated foam 10 ppi	800	94
Reticulated foam 40 ppi	3000	94
Reticulated foam 80 ppi	5800	94

One more factor must be considered before we summarize this foundation and discuss case histories: the factor concept that could explain why hydrophilic polyurethane has shown an improved efficiency in development of biofilms. In a reconsideration of the biofilm model (Figure 5.4), let us consider a model in which the substratum has reservoir capacity (Figure 5.5).

Without an impermeable wall behind a biofilm, carbon-containing pollutants would be absorbed. If the pollutant load varies from time to time, the wall could serve as a reservoir for excess pollutants. When the load is smaller, diffusion out to the layer could maintain a viable colony. Also, if low molecular nutrients (micronutrients, for example) are added artificially, they can be stored in the absorptive layer and released on demand.

BIOCHEMICAL TRANSFORMATION
OF WASTEWATER: SUMMARY

Chapter 4 described a more or less synthetic assault on the environment and the relationship of population and industrial development. A natural result of such activities is the release to the environment of certain synthetic chemicals including pesticides, solvents, and petrochemicals. We discussed the use of extraction techniques to mitigate the damage.

This chapter discusses the direct assault on the environment from humans: the contamination of water with organic matter, and describes the systems built to handle this type of pollution as biochemically based, that is, by culturing large engineering systems of various types with bacteria, the carbon-based pollutants are converted to CO_2 and biomass. The biomass is then separated, leaving the water free to reenter the natural environment.

Figure 4.3 shows the major parts of a recycle system. In practice, however, depending on municipal wastewater treatment systems to treat normal industrial effluents is impractical. The effluent may contain degradation-resistant chemicals

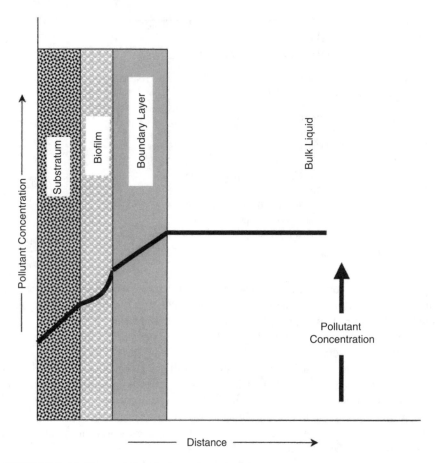

FIGURE 5.5 Biofilm model with absorptive substratum.

that are not properly treated by the typical wastewater system; the loads may be more than the system can handle. Most importantly, the cost of treatment is difficult to allocate equitably. If the costs are too high, the operation could become unprofitable. If the charges are too low, the public may have to bear the economic burden.

Most agriculturally based pollutants probably return to the environment at points that are difficult to identify and are called nonpoint source emissions. Pesticides leave farms as runoff or leach through the soil. Isolating and treating them is not often possible. While techniques such as ditching and basins are being adopted, the problem of treatment in the dilute state in which pollutants will be isolated is a serious concern.

The intent of Figure 4.3 is to show how polyurethane can serve as an important tool in the remediation of industrial and agricultural wastes. We will review the work of a host of researchers who recognized the benefits of polyurethane as a scaffold for colonization of pollutant-degrading organisms. While we focus on the treatment of water, bioreactors are also used to treat air emissions. The treatment of wastewater is well known to emit hydrogen sulfide to the environment. While the emissions act more as nuisances than health hazards, they nevertheless must be addressed. Later in this chapter, we will discuss the use of reticulated foam to deal with this problem.

In the case studies to follow, both hydrophilic and hydrophobic polyurethanes are used to affect remediation of polluted air and water. We will not discuss conventional wastewater systems because they represent large public works projects that dot the developed world. The first three case studies cover the use of reticulated foam as a scaffold for the remediation of polluted air. Another involves the use of a hydrophilic foam as a scaffold for a biofilter to treat aquaculture wastewater, permitting its return to the system (closing the recycle loop). Lastly, we will review our work on a composite of hydrophilic polyurethane grafted onto a reticulated foam to treat VOC-contaminated air.

CONVENTIONAL RETICULATED POLYURETHANE AS SCAFFOLD FOR MICROORGANISMS

While our focus has been on the treatment of wastewater, another technique known as biofiltering is intended to treat contaminated air. The most common technique and the one we will cite in the case studies is a subclass known as trickling filters. The general engineering design involves a countercurrent flow of air and water not unlike a scrubber. The exchange or extraction of the pollutant from the air into the water is the dominant step in the process. At steady state, the water is the vehicle within which remediation takes place. Thus while the pollution is resident in the air initially, the remediation takes place from the water and thus relates the process to the discussion above.

Devinny[46] reported on a biofilter system at a cigarette factory in Berlin, Germany. The intent was to reduce odor emissions to an acceptable level determined to be 10% of prebiofilter levels. To effect a reduction in odor, a trickling filter design was developed and the filter was installed on the roof of the facility. The footprint covered 250 m^2. The packing of the column was commercially available reticulated polyurethane marketed by Zander Umwelt GmbH, Nuremberg, Germany. The product was specifically designed for biofilter use and selected because of its weight. A volume of 500 m^3 was required, and the roof installation put constraints on the mass of the biofilter.

The Zander material was also chosen for its high surface area (600 m^2/m^3). The polyurethane was supplied in 4-cm cubes. The water phase was supplied intermittently and included a nutrient. Air flow through the system was 160,000 m^3/h, downward through the packing. The residence time in the biofilter was approximately 10 sec and the mean pressure drop was 400 Pa.

The inoculation method was not mentioned in the review, but it is safe to assume that the colony of microorganisms was developed by natural selection, that is, the air or water used in the operation was contaminated with microorganisms, as is true for all but sterilized fluids. The organisms were expected to attach and those that found sufficient food (based on their ability to metabolize the dominant pollutant in the air) would survive. It took 2 months to develop a colony of sufficient size to reduce the odor emissions at the tobacco factory by 90%. A consistent concern of biofilters is clogging due to the development of biomass. This is evidenced by a gradual increase in pressure drop across the bed. This was not seen in the project, and, in fact, the pressure drop was relatively stable.

In another project involving a reticulated foam from Zander, the air emissions from a wastewater treatment facility in Orange County, CA were treated.[47] Before the development of a biofilter, the Orange County Sanitary District (OCSD) treated the hydrogen sulfide and VOC emissions from its wastewater treatment facility by passing the air through a chemical scrubber. In this technique, the air passes through a packed column countercurrent to a spray of water moving downward through the column. The packing in the column was 3.5-in. Tri-Pack® (Jaeger Products, Inc., Houston, TX). Each of the two scrubbers constructed of fiber-reinforced polymer was 9.75 m high and 1.82 m in diameter (inside). The packing was in a 3.66-m section of the column with a head space above and a liquid reservoir below. Water was taken from the reservoir at the bottom of the column and pumped at 56.8 m^3/hr to the top. The air flow through each column was 16,300 m^3/hr.

The device was a chemical scrubber in which chemicals were added to the water passing counter-current to the air flow to react with the pollutants. Typical oxidizing agents used included hypochlorites, peroxides, and permanganates. Caustic, hydrogen peroxide, and bleach were all used at different times. The OCSD was interested in exploring conversion of the scrubbers to biofilters in order to reduce the operating costs and the costs of chemicals.

The odorous gases appeared to be consistent with other wastewater treatment systems. They were composed of organic and inorganic compounds. Hydrogen sulfide, ammonia, and reduced sulfur compounds are considered the most malodorous. The air also included VOCs, some of which were degradable.

A pilot plant study determined that the Jaeger Tri-Pack was insufficient. The hypothesis was that the relatively low surface area of the Tri-Pack compared to reticulated polyurethane was a contributing factor, and the colonization of the Tri-Pack material was low. Steady-state operation of the pilot plant was achieved in 5 days. After the pilot project was completed, full scale testing was performed using reticulated foam as the packing. With confidence developed with the pilot plant, the full-scale scrubber was converted to a biofilter using the Zander polyurethane as the packing or substratum for biological colonization. The inoculation of the scaffold was accomplished by filling the bottom reservoir with sludge from the wastewater treatment plant. The water was pumped over the packing for 24 h. Contaminated air was pumped through the column following the inoculation period. Reductions of H_2S concentration and pH were first observed on the third day of operation. By the ninth day, the reduction of H_2S reached 100%.

Several experiments were run to test the robustness of the system. Overall the performance was considered effective and the conclusions about the conversion of a chemical scrubber to a biofilter were positive.

Both of the above case studies used hydrophobic polyurethane as the substratum. The next study was intended to determine the effectiveness of a biofilter based on hydrophilic polyurethane cubes in permitting recycling of water in a closed-loop aquaculture facility.

USE OF HYDROPHILIC POLYURETHANE
IN AQUACULTURE

In the U.S., the customer demand for popular fish species such as striped bass (Morone saxatilis), trout (Salvelinus fontinalis), salmon (Salmo salar), tilapia, and catfish, (Clarinas gariepinus) stimulated research related to providing consistent supplies. The research led to the development of on-shore recirculating aquaculture systems (RASs). Such facilities rely on treating the outflow from the systems to remove pollutants and permit recycling of the water in compliance with local regulations on wastewater emissions and needs of certain fish species for temperature control. In some cases, sea water must be trucked from the ocean, and recycling is an economic necessity.

The contamination arose from fish waste and excess food and requires the elimination of proteinaceous materials, ammonia, and nitrates. Hydrophilic polyurethane was proposed as a substratum for a biofilter. The University of Maine sought to investigate the concept.[48] Two parallel 1700-l saltwater recycle systems were built. The biofilters were of the trickling filter design. Thirty Atlantic salmon were cultured and fed 0.5% average body weight diets. The effectiveness of a specially formulated hydrophilic polyurethane was compared to BioBall, a commercial biofilter packing material produced by Biotechnologies Frontiers (North Ryde, Australia).

An interesting aspect of these experiments was the entrapment of activated charcoal in the hydrophilic polyurethane matrix. Givens and Sacks[49] showed the potential of carbon-infused polyurethane foam as a filter medium. They showed a 99% removal efficiency of organics and nitrogenous compounds. While their study concerned domestic wastewater, the effect was assumed to be translatable to an aquaculture environment.

One of the problems encountered in the entrapment of activated charcoal in a polymer matrix is the "blinding-off" of the pores of the charcoal, thus inactivating it. Because the pores of the carbon are responsible for its ability to adsorb organics, any pores that are filled or coated with a polymer matrix reduce its effectiveness. Conventional treatments involve blending the carbon into an acrylic latex and then applying the slurry to a reticulated foam. Upon drying, a coalescence encapsulates the carbon.

The entrapment of an activated charcoal in a polyurethane follows a different mechanism. Since gelation occurs before drying, it is apparent that the pores are not fully blinded and upon gelation the pores are propped open.

The researchers on the aquaculture project decided to incorporate activated charcoal into the foam.[48] Procedures recommended by the manufacturer were followed in making the foam. The carbon was mixed in the aqueous phase, but the concentration was not given. It is therefore assumed that the carbon had some activity, but this was not estimated and therefore the experiments do not confirm that it had any effect on the efficiency of the material as a biofilter.

In the first of several experiments, the effectiveness of the hydrophilic polyurethane was compared to that of BioBalls. From previous work it was determined that comparable efficiencies of the two systems were found at a volumetric ratio of one

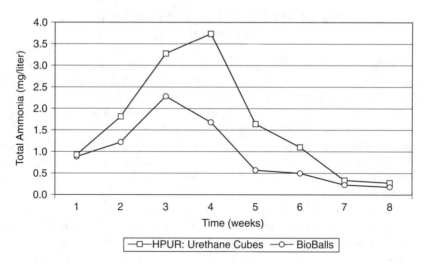

FIGURE 5.6 Performance of aquaculture recycle system.

part hydrophilic foam cut into 1-in. cubes to three parts BioBalls. Accordingly, the hydrophilic polyurethane experiments used a bed of 0.0246 m³ as compared to 0.0818 m³ of BioBalls. Inoculation was done with aquaculture wastewater. Atlantic salmon were placed in the culture unit and fed over a 3-week period after which feeding was stopped. Figure 5.6 shows the ammonia levels as a function of time. The conclusion was that the systems were comparable despite the volumetric differences in packing materials.

In another set of experiments, the wastewater was spiked with ammonium chloride and sodium nitrite. Within 4 days, the biofilters responded by increasing the biomass to metabolize the high nutrient load. The biofilters were challenged by increasing the fish population. The results allowed the author to calculate optimum biofilter size by ammonia load.

In an interesting conclusion, the author compared the results of the University of Maine study with studies in the literature using alternate packing materials. The other studies used 8.9-mm Koch rings[50] and 5-cm limestone. Table 5.3 shows the efficiencies of biofilters normalized to biomass density.

TABLE 5.3
Volumetric Efficiencies of Biofilter Packing Materials

Column Packing	Weight of Fish/Volume of Packing Material (kg/m³)
Koch rings	128
Limestone	70
Hydrophilic polyurethane	583

An important point must be made concerning the colonies of microorganisms on the foam. It is implicit from the data that cutting the foam into cubes increases the effectiveness of the filter. It is logical, therefore, that mass transport through the foam is not the mechanism at work. Rather, the flow around the cubes provides the contact necessary for the metabolism to proceed. Other studies confirmed that colonization occurs on the surfaces of the cubes or in the large pores.[5] It is clear that this limits the effectiveness of the polyurethane by reducing the potentially large internal surface area to the surface of a cube. Despite this limitation, the polyurethane cubes showed volumetric advantages over the polyolefin BioBall spheres, and this limitation led to our research in grafting a hydrophilic polyurethane coating to the inside surface of a flow-through reticulated polyurethane.[14]

USE OF HYDROPHILIC–HYDROPHOBIC COMPOSITE IN AIR BIOFILTER

The effectiveness of this technology was reported in a study we sponsored at the University of California, Riverside, under the direction of Professor M.A. Deschusse.[51] While the overall project was unsuccessful, the ability to support biological growth in a flow-through structure was confirmed. The effectiveness of the composite compared to a combination of hydrophobic polyurethane and the Zander material showed the method to have great potential.

Using the techniques discussed in Chapter 3, an MDI-based hydrophilic poly-urethane (Chemron Corporation, Paso Robles, CA) was grafted to a 30-ppi polyether polyurethane reticulated foam (Crest T-30). The object of the study was to investigate the products as packing materials for a biofilter to remove toluene and H_2S from air. Performances of biofilters and biotrickling filters are typically reported as elimination capacity vs. load. Figure 5.7 shows hypothetical characteristics and defines critical loading — the highest loading at which near-complete removal is observed.

Four laboratory-scale biotrickling filters were constructed. Two were used for the treatment of H_2S. A control filter was packed with untreated reticulated foam and the other was packed with the grafted composite. Two additional filters (one control, one composite) were used for toluene treatment. The biotrickling filters were made of clear polyvinyl chloride pipe (15.4 cm inner diameter; 121.9 cm total bed height). Each biotrickling filter was filled with foam for a bed volume of 23.6 l. Figure 5.8 and Figure 5.9 show the experimental setup. Figure 5.10 is a photograph of the pilot plant.

Serious problems occurred with the foam. The biomass development was of sufficient weight to collapse the foam. The collapse was sufficiently serious to end the experiments prematurely (after 28 days). The problem was particularly serious in the toluene-degrading biotrickling filters which, by nature of the process, had the greatest biomass growth. In the H_2S-degrading biotrickling filters, the biomass growth was virtually nil and bed stability was good.

This experiment illustrates the interdependence of chemistry and engineering. The attempt to improve the surface on which the microorganisms were to grow was pursued without regard for the strength of the materials used. The polyether reticulated

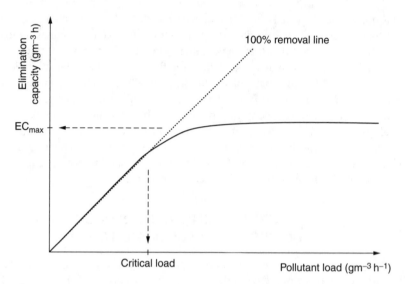

FIGURE 5.7 Typical elimination capacity-vs.-load characteristics of a biofilter. EC_{max} is the maximum elimination capacity. Critical load is the maximum loading at which the removal efficiency starts to deviate significantly from the 100% removal line.

foam was chosen for its mass transport and surface area characteristics. Little consideration was given to the factor that led to the failure of the project: compressive strength. The activity of the reticulated foam was at least adequate and arguably better than those of the materials to which it was compared. However, failing to consider the engineering stresses to which the material would be subjected meant it had to be considered unacceptable. Nevertheless, the project developed sufficient data to lead to the optimistic conclusion that a biofilter based on the composite would represent an advance in the technology. Table 5.4 summarizes the conditions of the pilot project.

As a result of the collapsing foam, two grades of grafted polyurethane were tested in the toluene study. The first (Composite 1) was produced without regard to compressive strength. It lasted 16 days in operation until the experiment had to be terminated. To remedy the problem and allow the experiment to continue, a technique was developed to stiffen the same reticulated foam without materially affecting its hydrophilicity. The material represented a great improvement, lasting almost 60 days. It too, however, ultimately collapsed under the weight of the biomass. Before collapse, however, sufficient data were gathered to show a consistent difference between the hydrophilic polyurethane-grafted reticulated foam and the reticulated foam. Figure 5.11 shows the comparison.

While the primary purpose of the study was to confirm the advantage of grafting a hydrophilic surface to a flow-through scaffold, we were also able to develop a comparison with the Zander product. We should note that the packing of the Zander product may not be optimum. Inasmuch as it is a reticulated polyurethane and compares quantitatively to the reticulated foam used for this study, we have some confidence that the differences observed are valid. Figure 5.12 illustrates the comparison of the

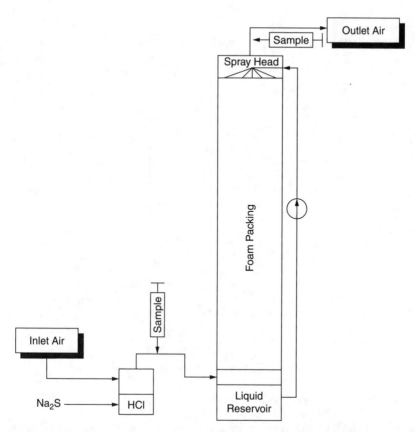

FIGURE 5.8 Countercurrent pilot plant for remediation of H₂S-contaminated air.

steady-state performances of the various packing materials. In order to achieve representative performance without the collapse of the foam, certain adjustments were made. The composite foam curve fit is the combination of the two grades. Data for the softer foam were taken from Days 0 through 18. Data from the stiffer foam were obtained from Days 14 through 56. The reticulated foam data were taken from 0 to 72 days; the Zander foam data were obtained between Days 14 and 38.

While the overall study revealed engineering deficiencies, the chemistry appeared to offer certain advantages. This study did not address the reasons for the efficiency improvement. We believe the improvement arose from the combination of a more biocompatible hydrophilic surface and the buffering of the absorbent substratum discussed earlier.

OTHER PROJECTS

Fava et al.[52] developed a bioreactor to degrade chlorinated hydrocarbons in waste-water. Among other materials, polyurethane cubes were used as scaffolds. The degrading cells were immobilized in a fixed-bed reactor and showed significant

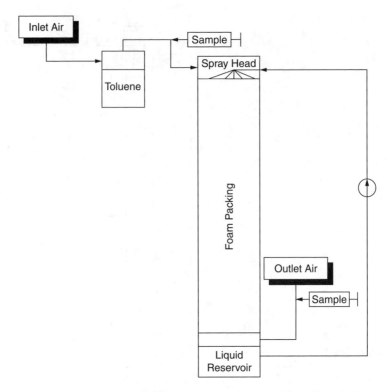

FIGURE 5.9 Co-current pilot plant for remediation of toluene-contaminated air.

activity in degrading dichlorobiphenyls to chlorobenzoic acid and chloride ions. A ring opening mechanism was proposed. The distribution of cells in the bioreactor was evaluated using an electron microscope and the kinetics of the conversion are discussed.

Zaiat studied a horizontal flow bioreactor based on polyurethane foam in which was immobilized a sludge from swine wastewater.[53] The bioreactor was operated anaerobically. Two porosities were investigated in an effort to define flow characteristics in combination with degrading efficiency. Glucose was selected as the main carbon source. The chemical oxygen demand (COD, a measure of the amount of degradable carbon) was reduced by 96%. The efficiency of the bioreactor, however, was found to be a function of porosity. This was explained by the development of channeling and short-circuited flow through the bed since pieces of foam were used instead of an integral block. In another study, Zaiat investigated superficial velocity through a similar reactor.[54] As expected, the velocity through the bed (which determines residence time) affected the conversion.

Sun et al. studied the immobilization of *Rhizopus orzae* in polyurethane foam cubes.[55] The growth rates of the cells, the effectiveness of the immobilization process, and the ability to produce lactic acid were investigated. Although the study was not an attempt to remediate a wastewater stream, the paper presents an excellent review of the surface effects that hampered the development. The size of the cube defines

FIGURE 5.10 Pilot plant at University of California, Riverside.

the rate. This suggests that the surface areas of the polyurethane foam cubes define the temporal efficiency and also that mass transport inside the foam cube is minimal. We noted these characteristics in the case studies above, specifically the aquaculture recycle project.

Borja and Banks studied the kinetics of the anaerobic digestion of fruit processing wastewater (COD = 5.1 g/l).[56] They used several substrata as scaffolds for the development of degrading colonies of bacteria. The kinetics of degradation were compared with the kinetics of a suspended biomass. The first order rate constant was determined and is shown in Figure 5.13. While the rate for the polyurethane was lower than rates for the sepiolite and saponite, some of the improvements we suggested and an improvement in mass transport could improve the rates of reaction. In any case, the paper of Borja and Banks suggests effect on kinetics exerted by the substratum.

Borja and Banks also performed a kinetic study of the anaerobic digestion of soft drink wastewater using bioreactors containing various suspended supports. Again, bentonite, zeolite, sepiolite, saponite, and polyurethane foam were used as substrata onto which the microorganisms affecting purification were immobilized. Similar results were noted. Vieira et al. also studied the kinetics of anaerobic sludge immobilization in the matrix of a polyurethane foam.[57]

Cell washout from nonadhering polyurethane has been both a problem and a benefit. While a strong bond between the substratum and the cells is a requirement for certain product designs, the weak bond developed between a hydrophilic cell

TABLE 5.4
Biotrickling Filter Characteristics

Characteristic	Toluene Biotrickling Filters	H₂S Biotrickling Filters
Bed height and internal diameter (cm)	121.9 × 15.7	121.9 × 15.7
Bed volume (l)	23.6	23.6
Packing height of reticulated foam reactor	130 × 6″ (L × H, 8 rolls), 598.8 g	130 × 6″ (L × H, 8 rolls), 619.9 g
Packing height of composite foam reactor	130 × 5″ (L × H, 9 rolls) + 130 × 3″ (1 roll), 763.4 g	130 × 5″ (L × H, 9 rolls) + 130 × 3″ (1 roll), 813.4 g
Recycle liquid volume (l)	4	4
Gas–liquid flow	Cocurrent	Countercurrent
Gas flow rate, EBRT[a]	1.4 to 2.8 m³/h, 60 to 30 sec	1.4 to 2.8 m³/h, 60 to 30 sec
Liquid trickling rate	10 gph (37.8 l/h)	10 gph (37.8 l/h) followed by switch to 1.5 to 5 gph
Operating temperature (°C)	19 to 20	19 to 20
Pollutant inlet concentration	0.4 to 1 g/m³	Variable, up to 230 ppm
Medium feed rate (l/d)	6 to 7	2.4
Medium nature	Mineral	OCSD secondary effluent
Medium feed pH	7 to 8	7 to 8
Operating pH	6 to 8	1.6 to 2

[a] EBRT (empty bed [gas] residence time) = bed volume/air flow rate.

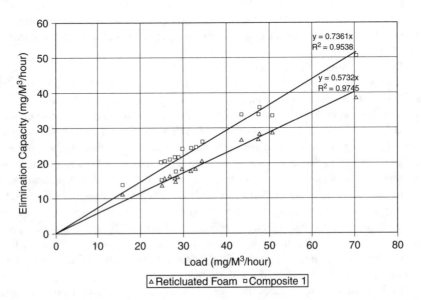

FIGURE 5.11 Comparison of elimination capacities of reticulated foam and hydrophilic polyurethane-grafted reticulate.

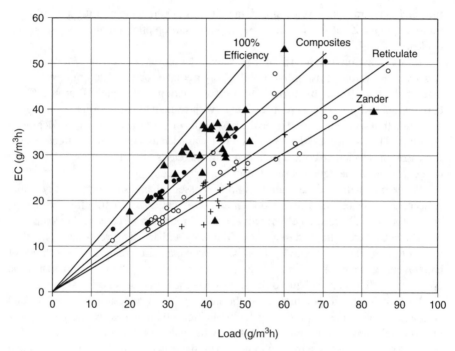

FIGURE 5.12 Comparison of packing materials.

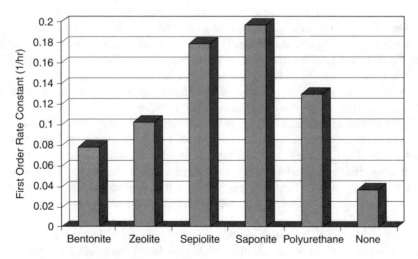

FIGURE 5.13 Efficacies of various substrata.

wall and hydrophobic polyurethane surface causes the cells to slough off. For biomass control, the ability to wash out cells, thereby maintaining some level mass transport through the body of the foam, is considered an essential part of the design. Zaiat et al. studied this feature in an anaerobic packed bed reactor in which an activated sludge was immobilized.[54] They investigated the optimization of the shear produced by water flow and mass transport. As we noted, the kinetics in part serve as a function of mass transport. In the Zaiat study, a decrease in pressure drop after a high flow period was evidence of wash-out.

Immobilization of the trichloroethylene-degrading bacterium *Burkholderia cepacia* was evaluated using hydrophilic polyurethane foam.[58] The organisms were dispersed in the foam by using various formulation techniques. Pluronic block copolymers and silicones were investigated with respect to their ability to entrap the cells. Lecithin was determined to be the best performer. While the organism degraded chlorinated hydrocarbons, the respiration rate was used as a measure of effectiveness. Increases in surfactant concentration led to increases in cell washout. High cell concentration in the aqueous resulted in low washout.

While the quality of the foam was not discussed, changes in surfactant type and concentration were the primary determinants of cell size, distribution, and type and doubtless affected the cell effectiveness and retention of cells in the foams.

The cells were entrapped in a hydrophilic polyurethane. This represents one technique of immobilization. With hydrophilic polyurethanes, a researcher has a choice of entrapment or adhesion. With adhesion, a broth containing the cells inoculates the polyurethane. The cells adhere to the surface via their natural adhesive systems.

With hydrophobic polyurethanes, the only option is adhesion. The polyol and isocyanate environments producing hydrophobic polyurethanes by the prepolymer method are thought to be too severe for living cells. As noted earlier, the weaker adhesion of adsorption to a hydrophobic polyurethane can be an advantage. Sanroman compared the adsorption and entrapment techniques[59] and determined that the adsorption technique was superior based on citric acid productivity and operational stability.

Our discussion of cells in an HPU12 may shed some light on this. Entrapping the cells in a hydrophilic polyurethane certainly affects the diffusion of substrate into the domains of the cells. An interesting study would involve hydrogels with high equilibrium moisture levels.

Nemati investigated the inhibiting effect of ferrous iron on the rate of oxidation in a bioreactor packed with a polyurethane foam.[60] Fynn and Whitmore investigated the ability of methanogen species to colonize reticulated polyurethane foam in a continuous culture system. Electron micrographs confirmed that two methanogen species colonized the matrix.[61] The methane output was superior to a liquid culture control.

Rao and Hall compared the activities of algae, cyanobacteria, and photosynthetic bacteria entrapped in an alginate gel and in a hydrophilic polyurethane.[62] Their opinion was that such systems could maintain effectiveness for years using sunlight as an energy source.

Perhaps the most interesting application of polyurethane foam as a substratum for cell growth was studied by Bailliez et al.[63] While not specifically a remediation study, their work compared hydrophobic and hydrophilic polyurethanes, TDI- and MDI-based prepolymers, and entrapment and adsorption methods, and also investigated the production of hydrocarbons by *Botryococcus braunii*. An unfortunate feature of biotechnical research in the use of polyurethanes is that the chemistry is rarely explained. While Bailliez includes some detail, much of their work simply designates products without specific references to the polyols. It is, of course, part of the mission of this book to show that polyurethanes are specialty chemicals. It cannot be assumed that the only differentiation is whether they are hydrophilic or hydrophobic. The most important differentiations are diffusion through a membrane polyurethane and the hard and soft segmentations.

Nevertheless, Bailliez made progress toward full disclosure by differentiating the 11 prepolymers his group tested. According to the text, they were polyethers of polypropylene glycol and copolymers of polypropylene glycol and polyethylene glucol. The mix of prepolymers included at least three hydrophilics (Hypol 2002, 4000, and 5000). All of the prepolymers except Hypol 4000 and 5000 were TDI-based. Hypol 4000 and 5000 are MDI based.

In the entrapment studies, a concentrated dispersion of the cells was mixed with an equal mass of prepolymer. For adsorption, the foams were prepared, washed, sterilized in an autoclave, then placed in suspensions of the algae for 14 days. The entrapped algae showed no respiration for the majority of the prepolymers including the TDI-based hydrophilic Hypol 2002. Only five of the prepolymers were thought to be acceptable; they included the two MDI-based prepolymers (Hypol 4000 and 5000).

The polyurethane-entrapped algae showed reasonable activity compared to free suspensions of cells. The activity fell sharply after a few days, however. The authors hypothesized that the decrease could have been the result of residual toxicity of the polyurethanes. They also noted that the previous method of entrapping the cells in alginate gels performed better. Our work focused on the differences in diffusion through the entrapping polymer. Baillez et al. also discussed mass transport in the foams as a contributing factor. In summary, direct entrapment resulted in an immediate reduction in viability of the cells.

Adsorption of the cells on the polyurethanes was more successful. The research focused on the two MDI-based prepolymers. In the inoculation phase, an immobilization yield of 70% was achieved. It should be noted that immobilization was observed in the large pores only. Hypol 5000 was thought to be the better of the two MDI-based prepolymers, but the reason was not hypothesized. While the production rates were lower than those of the free suspensions, the production was sufficient to be considered a viable technique. Table 5.5 presents the data.

TABLE 5.5
Production of Hydrocarbons by Immobilized Algae

System	Biomass (grams/liter)	Hydrocarbon Production (grams/liter)
Control	3.4	0.73
Hypol 5000	2.74	0.44
Hypol 4000	2.14	0.41

6 Biomedical Applications of Polyurethane

We have discussed the use of polyurethanes in environmental applications. Certainly the conditions under which a polyurethane must operate are by no means mild. Arguably, however, the most severe environment one can imagine in which a material must operate and above all cause no harm is inside the human body. Not only must the material function in above-ambient temperatures, but it is continuously held in a hydrolytic environment containing chemicals and proteins. Furthermore, the human body has developed defensive strategies that attack foreign bodies by attempting to degrade or encapsulate them. A polyurethane product designer who develops a device intended for implantation into the human body must consider additional stresses not normally encountered in more traditional applications.

In this chapter and the one that follows, we will review the research concerning the uses of polyurethanes in biomedical applications. Such uses range from topical application of hydrophilic polyurethane pads to the implantation of scaffolds of reticulated foam as organ-assist devices. The range of applications is broad and each use requires that specific problems associated be addressed. This chapter begins with a discussion of biocompatibility — a broad concept ranging from the simple nonirritating characteristics required for topical applications to the complex type of compatibility (hemocompatibility) that allows use with whole blood.

As we will show, biocompatibility of a device can be approached from two perspectives. This is most clearly established in a discussion of how blood or plasma interacts with a foreign body. We can approach a project with the goal of allowing the body to react to the device and encapsulate it by a natural process, or alternatively, adopt a strategy to make the material as transparent as possible. Both techniques have been used with some degree of success. It is testimony to the versatility of polyurethanes that either approach can be accomplished with a simple change in the polyol.

Once a baseline of biocompatibility has been established, we can discuss specific design features. *In vivo* biodegradability is of current interest with regard to pseudo-topical applications (burn dressings) and advanced medical devices (organ-assist devices). We will also discuss cell attachment inasmuch as certain attachment-dependent cells cannot survive without association with a surface (synthetic or natural). It is interesting to note that this area of biological science is in a rapid state of flux. Ten years ago, cell attachment was discussed in qualitative terms. Current research focuses on antibody–antigen reactions and notch–ligand associations. Whatever mechanism is chosen, we know that in order for a polyurethane to function as a cell attachment system (a scaffold, for example), provisions must be made for the cells to bond to the material.

TABLE 6.1
Tests Required for Contact with Breached
Surface for 28 Days

1	Primary skin irritation
2	Cytotoxicity: agar diffusion test
3	Hemolysis: rabbit blood
4	Kligman maximization study (sodium chloride extract)
5	Systemic injection test
6	14-day repeated intravenous (subchronic) toxicity study

BIOCOMPATIBILITY

Biocompatibility for topical devices is typically limited to determine whether a material causes sensitization. It has become customary to begin studies of topical biocompatibility by determining whether the material is cytotoxic. We say this not from the perspective of a toxicologist, but as a practical matter. Cytotoxicity is a pretty good indicator of general compatibility. Testing cytotoxicity is relatively inexpensive. If a material fails a cytotoxicity test, it has little chance of passing other tests. If it passes, a researcher can proceed with some level of confidence that the material will pass other tests. However, too many contradictory test results make it impossible to ignore other indicators of compatibility. In fact, the International Organization for Standardization (ISO) and U.S. Food and Drug Administration (FDA) require six tests to confirm the safety of a device in contact with a breached surface (typical wound dressing) for a duration of 28 days. Table 6.1 lists the tests.

Separating the individual tests by cost, we devised a protocol that is useful as a safety screen (Figure 6.1). In effect, it postpones the more expensive tests until a material passes the less expensive tests.

In our experience, polyurethanes have proven safe for topical applications. Extractable formulations including surfactants, catalysts, chain extenders, and other components used in manufacturing devices are more problematic and cause the most concern. Furthermore, all toluene diisocyante (TDI)-based polyurethanes are subject to hydrolysis that produces small amounts of toluene diamine, which has been confirmed as a carcinogen. Interesting enough, there is a strong distinction between this confirmed chronic toxicity and the suspected toxicities of MDI-based polyurethanes.

This subject was studied in relation to the silicone breast implant controversy of the 1990s. Silicone-filled artificial breasts were coated with thin layers of a TDI-based polyurethane foam to encourage stabilization by cell ingrowth. Residual toluene diamine (TDA) was found in prepolymer-based polyurethanes at the part-per-billion level and, more importantly, hydrolysis that would lead to the release of more TDA *in vivo* was suspected.

A panel of researchers at the FDA studied both effects and developed procedures to calculate the chronic risk. The confirmation of the risk, albeit very small, led to special attention rightfully paid to implantable devices based on TDI. Due in part to the long history of implanted polyurethanes based on MDI, such polyurethanes

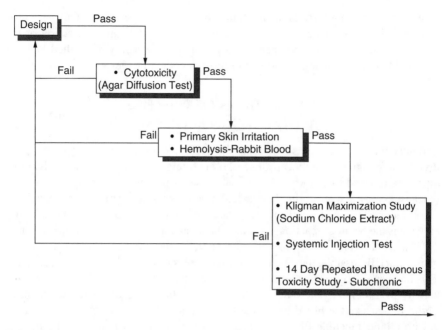

FIGURE 6.1 Cost-effective testing protocol for establishing safety of a wound dressing.

are not subjected to the same special attention. Sometimes biocompatibility is a perceived quality.

More problematic than topical devices, however, are devices that penetrate the dermis and are therefore rightfully deemed implantable. This category includes devices used to treat third-degree burns. In current practice, the affected area is excised and a dressing is placed within the wound and perhaps sutured. The dressing is considered an implant from a device-safety perspective.

Biocompatibility of implanted devices is exponentially more complicated than biocompatibility of topical devices. Rejection of a topical device is not a major issue but it represents the primary difficulty for a device that enters the body. The device will come into contact with fluids and cells that are sensitive to contacts with foreign materials. The success of the biofilters discussed in the last chapters is an implication of a level of biocompatibility. Bacteria are somewhat forgiving because they are able to tolerate contacts that would cause instant death to mammalian cells. Bacteria have evolved to the point where they are able to attach themselves to almost anything.

Mammalian physiology, particularly *in vivo*, took the opposite evolutionary course. Mammals developed defenses that "attack" foreign materials, and these defenses pose the greatest difficulties a biomedical designer has to face. If not for this tendency to attack foreign materials, the development of remedial devices such as artificial heart pumps and lungs would simply follow the techniques for designing classic chemical engineering projects.

Exposing polymers to a biological environment elicits a number of reactions such as protein adsorption and cell adhesion. In some cases, the actions of body defenses against a material are planned as part of a strategy. In other situations, the strategy is to make a material as neutral as possible. Deficiencies in either strategy can result in catastrophic failure of a device. In both situations, the action of the body upon the device must be understood, and more importantly, controlled as much as possible. The following sections will discuss both strategies.

INTERACTIONS OF PROTEINS WITH FOREIGN MATERIALS

The attachments of proteins and cells to a surface represent the body's attempt to dissolve a foreign body or encapsulate it. If the foreign body is in contact with blood, a process referred to as intrinsic coagulation seeks to isolate the body by covering (encapsulating) it with fibrin. The coagulation process is a cascade of interrelated blood factors that work sequentially to accomplish encapsulation. The coagulation cascade is the result of reactions of individual coagulation factors listed in Table 6.2.

The cascade process begins with "detection" of the surface by Factor XII (Hageman factor). If possible, the enzyme adheres to the surface and is said to be activated. This has been described as an opening of the protein to expose additional enzyme activity in the blood. The other factors follow, culminating in the development of fibrin. In a sense, the blood has made the foreign body biocompatible by isolating it with a biocompatible fibrin coating. Unfortunately, this activity can destroy the function of the device. For example, with a vascular graft, the fibrin capsule could inhibit blood flow through the vessel.

One strategy for interrupting the coagulation cascade is to inhibit the adsorption of protein to the surface of the device. The possibility of accomplishing this is hinted at in the experiment described below. Consider the following data comparing the adsorption of albumin on various surfaces. The materials were washed with a 1-mg/ml solution of albumin until a steady state was achieved.

TABLE 6.2
Blood Coagulation Factors

Factor	Common Name
I	Fibrinogen
II	Prothrombin
III	Tissue thromboplastin
IV	Calcium
V	Proaccelerin
VI	Omitted
VII	Proconvertin
VIII	Antihemophilic factor
IX	Plasma thromboplastin
X	Stuart–Power factor
XI	Plasma thromboplastin antecedent
XII	Hageman factor
XIII	Fibrin stabilizing factor

TABLE 6.3
Adsorption of Albumin on Various Surfaces

Surface	Adsorption ($\mu g/ml^{2)}$)
Hydrophilic polyurethane 1	0.020
Hydrophilic polyurethane 2	0.037
Collagen	0.086
Siliconized glass	0.176
Polystyrene	0.196
Hydrophobic polyurethane	0.571

Adapted from Brash, J.L., and Uniyal, S.[64]

It appears from these data that adsorption to hydrophilic surfaces would be a viable strategy. Other research produced a similar conclusion. Waugh et al.[65] showed that prothrombin (another member of the coagulation cascade) binds more strongly to polymethyl methacrylate than glass. Chaung[66] showed that plasma proteins bind more strongly to polyvinyl chloride than a hydrophilic dialysis membrane.

This is not the only approach to biocompatibility, however, and is, in fact, not even the most common approach. Two major approaches are commonly used. The technique cited above seeks to be nonthrombogenic (referring to the last stage of the coagulation cascade, the reaction of thrombin and fibrinogen to produce the fibrin capsule). The alternative approach to biocompatibility involves the development of a biocompatible or passivating layer. The use of Dacron for vascular graft is an example utilizing this technique.

A biologic surface that develops an endothelial cell surface is referred to as a neointima. If it is covered with blood components such as fibrin, it is called a pseudointima. In both cases, the surfaces are passive with respect to the blood to which they come into contact. A pseudointima, however, is typically unstable and subject to further thrombic response. If the surface is damaged, as during surgical implantation, a catastrophic failure can result. This coupled with the difficulty of developing a complete endothelial layer caused one researcher to describe a device as "physiologically tolerable" rather than biocompatible or hemocompatible.

The factor that differentiates natural and synthetic surfaces is the endothelial layer. The layer is truly nonthrombogenic and is capable of repairing itself. While it can develop on synthetic material, its growth in humans is slow. One way of accelerating growth is to seed the synthetic material.

To develop the endothelium, provision must be made for attachment of the cells. This attachment can be accomplished by connection to a fibrin capsule, physical attachment to a rough surface (as in a bundle of fibers) or via adsorbed or covalently bonded antigen ligands. We will expand on these concepts (development of endothelium and antithrombic hydrophilic polymers), but it is useful to mention that both strategies are possible with properly formulated polyurethanes. Nonthrombic polyurethanes can conceivably be devised by hydrophilic polyurethanes or by the attachment of antithrombic ligands such as heparin or hirudin, both of which are conveniently attached with appropriate polyurethane prepolymers. Depending on the composition of the polyurethane, protein adsorption and cell attachment can be controlled. The special physical chemistry of a polyurethane gives it the physical properties required to handle the mechanical stresses required to meet design requirements and the formulation flexibility to resist fibrin development or encourage cell attachment. The properties depend on the polyol, and to a degree the physical form (foam or elastomer) selected.

The coagulation cascade is at first a protein adsorption phenomenon. One approach we suggested was to make the surface sufficiently hydrophilic to prevent adsorption of the proteins. This is not always possible, however. If the goal of research is to devise a vascular graft, certain physical requirements preclude hydrophilic polymers. An alternative would be a composite material, but that is problematic. Another approach — the one most researchers have followed — is interrupting the coagulation cascade at a point further along the process.

Strategies depend on the requirements of the device. The strength of the material is an issue. If the material is required to have some flexibility and physical strength, a hydrophobic polymer is required, but avoidance of Factor XII activation may not be convenient. In that case, the most common strategy is to interrupt the thrombin/fibrinogen reaction. In one technique, recombinant hirudin (an anticoagulant derived from leeches) is used to inactivate thrombin.[67] No matter what technique is used, a device in contact with blood or plasma requires a strategy to interrupt the coagulation cascade.

Coagulation is not the only problem with materials intended for implantation, however. Cardiac pacemakers are intended to correct arrhythmias. Insulating materials for a pacemaker lead must be tough and long lasting. The first leads were insulated with polyethylene or silicone rubber. Neither material was considered ideal because of endocardial reactions (polyethylene) and limited durability (silicone rubber). The strength and flexibility of polyurethanes led to their introduction in 1978 as lead insulators.

Pellethane is an extrudable thermoplastic polyurethane based on MDI and a hydrophobic polyol was used in this application. For the most part, the application was successful, but evidence of stress cracking became apparent. The combination of the hydrolytic environment and the polymer under tension caused failures that led to current leakage and ultimate failure of the device. As a result, softer grades of Pellethane and alternative annealing procedures were adopted and reduced the problem dramatically.

Despite the improvements, a more durable elastomer was clearly needed. A battery of polyurethane elastomers including Pellethane were prepared and implanted subcutaneously in rats. Before implantation, the polymers were extruded into tubes and elongated over mandrels to 400%. The implants were left in place for 6 months and examined under a scanning electron microscope. One of the better performing polyurethanes was based on poly(1,6-hexyl 1,2-ethyl carbonate) diol. Polyesters typically are considered less durable due to the presence of esterase enzymes *in vivo*. From the data recovered during the implant period, it was determined

TABLE 6.4
Commercially Available Medical Grade Polyurethanes

Trade Name	Manufacturer	Composition[a]	Comment
Biomer	Ethicon, Summerville, NJ	MDI–PTMEG–EDA	Thermoplastic elastomer
Corethane	Corvita, Miami, FL	MDI–carbonate–BD	Thermoplastic elastomer
Chronoflex	CardioTech, Woburn, MA	HMDI–carbonate–BD	Thermoplastic elastomer
Pellethane	Dow Chemical, LaPort, TX	MDI–PTMEG–BD	Thermoplastic elastomer
Tecoflex	Thermedics, Woburn, MA	HMDI–PTMEG–BD	Thermoplastic elastomer
Tecothane	Thermedics, Woburn, MA	MDI–PTMEG–BD	Thermoplastic elastomer

[a] MDI = methylene bis-diphenyldiisocyanate. HMDI = hydrogenated MDI. PTMEG = polytetrametyl ethylene glycol. EDA = ethylene diamine. BD = butanediol; carbonate is a proprietary carbonate diol.

that polyethers were subject to stress-cracking phenomena. Although the polycarbonate polyurethane is a polyester, the ester linkage is sufficiently hindered so as to prevent enzymatic degradation.[68] Table 6.4 lists a number of polyurethanes that are or were commercially available for medical devices.

AVOIDING COAGULATION CASCADE

When strength-of-materials considerations are not significant design features of an anticipated device, one can use strategies that seek to prevent the adsorption of proteins and cells. As Table 6.1 illustrated, a strategy to accomplish this would be to use hydrophilic polymers. The interaction of proteins and surfaces is a complex subject and depends on the nature of the protein, associations with other proteins, time, shear, and other factors including the chemistry of the surface.

We have often mentioned polyethylene glycols (PEGs) related to their roles as building blocks for polyurethanes. These polymers are also well known for their protein compatibility.[69] Adsorption was shown to be inversely related to the length of the PEG molecule. This work supported the work of Jeon discussed in Chapter 2.

Braatz et al.[70] studied the incorporation of PEG backbones into polyurethanes prepared with low residual isocyanate contents. The reason for this was that the intent of the study was to produce dense hydrogels rather than foams. Along with low isocyanate content, the researchers used low hard segment percentages. You will recall that the hard segments are also the hydrophobic segments of molecules, so the resultant polymers have higher equilibrium moisture potentials. The polymers Braatz used had 85% and higher equilibrium moistures. Of special interest was the use of isopherone diisocyanate. Several PEG and PPG polyols were used. The prepolymers were prepared according to the procedures discussed in Chapter 2. In the case of what Braatz referred to as Biopol XP-5, a triol of a random PEG–PPG polyol of 22,000 molecular weight was used.[71]

In a set of parallel studies, a series of polyethylene glycols were tested to confirm the effects of polyol molecular weight. The polymers including a TDI-based prepolymer

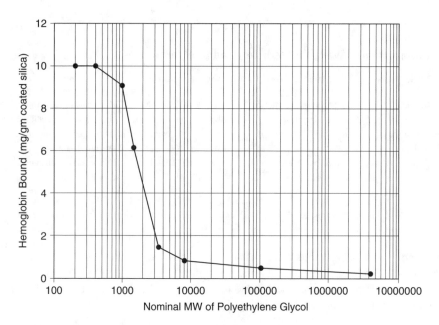

FIGURE 6.2 Adsorption of hemoglobin on ethylene glycol-coated silica.

using PEG 1000 as the polyol were coated onto silica for testing. The samples were then mixed with bovine hemoglobin. Protein adsorption was determined by dye adsorption and other methods. We report the dye adsorption method here.

The significance of this study goes to the heart of our task. In our discussion of the extraction of pollutants, we suggested that polyglycols had utility as extracting solvents, but because of their physical nature (water solubility) they were not useful. We proposed to polymerize the glycols into water-insoluble polymers by reactions with polyisocyanates. We then presented data to support the notion that the polymers maintained the solvent properties, but they were translated into a water-insoluble matrix (a polyurethane).

The analogy can be carried over to this example. The first set of data, supported by other researchers, shows that an ethylene glycol and copolymers of ethylene glycol and propylene glycol have minimal protein adsorption properties. They would qualify for consideration as biocompatibility factors by reducing the tendency to initiate the inflammation cascade. Figure 6.2 depicts the adsorption of hemoglobin on silica coated with ethylene glycols of increasing molecular weights.

Higher molecular weight glycols have no structural integrity; they are water-soluble waxes that are of no value without modification. The strategy is again to take the usable backbone and translate it into a useful form by reacting the material with a polyisocyanate. Table 6.5 describes prepolymers made according to the method taught in the product literature for high-performance isocyanates.[9] Along with changes in molecular weight are changes in the ratios of ethylene oxide (EO) and polyethylene oxide (PO). Surprisingly, the molecular weight had a more pronounced effect on adsorption. Figure 6.3 reports the data in graphic form.

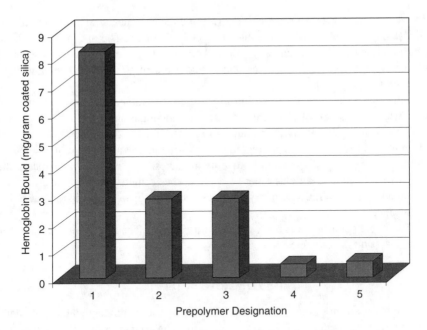

FIGURE 6.3 Effect of polyol molecular weight on the adsorption of proteins.

TABLE 6.5
Prepolymer Preparation

Prepolymer	PEO:PPO Ratio	MW	Hemoglobin Bound (mg/g coated silica)
1	0.15	1100	8.3
2	0.33	3438	2.9
3	8	2917	2.9
4	3	7000	0.5
5	3	22000	0.6

We have focused our attention on minimizing adsorption, but readers will note that the above data are still useful even if adsorption is the preferred technique. Braatz tested a polypropylene oxide polymer as part of his study and found that the protein adsorption was 6.4 mg/gram. Compare this with the data in Table 6.5.

SUMMARY

We have described biocompatibility from two perspectives. First, we described a strategy that would intentionally cause a coagulation cascade. We touched briefly on the environmental stresses inherent in implantable devices and the work done by several researchers to minimize that effect by the use of an alternate polyol. When a strong resilient polymer is a requirement for an anticipated device, preventing the coagulation cascade is not practical. Interrupting the cascade at the thrombin–fibrin-ogen stage was thought to be a viable strategy in that case. The use of immobilized heparin or ligand was mentioned. We described a strategy to minimize the adsorption of proteins, thus preventing the cascade by the use of high molecular weight poly-glycol-based polyurethanes.

This chapter is necessarily an introduction to this subject. It is our intention to introduce the strategy of affecting changes in biocompatibility by what you now know to be conventional polyurethane disciplines. In this discussion we have not differentiated the strategies with regard to the form the polyurethane will assume — another degree of freedom unique to polyurethanes. If a polymer that meets the design specification is an elastomer or a foam, the formulation difference is small compared to the overall chemistry changes required to improve biocompatibility, for instance.

Two more subjects must be covered in this chapter. The first is biodegradability and the second is cell adhesion.

BIODEGRADABILITY

Although we are not great proponents of medical use of biodegradable polymers, our preference should not influence readers because our primary task is to teach product design principles. It is fairly easy to make a biodegradable polyurethane. Although many scientists would disagree, the opinion is nevertheless based on sound principles. Researchers around the world study biodegradable polymers in efforts to develop scaffolds upon which mammalian cells will congregate. Expected applica-tions range from the repair of damaged cartilage to organ-assist devices.

For the most part the polymers exist in the forms of foams of varying quality. Elastomers are of interest as well. In our opinion, a polymer represents a classic case of optimizing chemistry and physical properties. The task is made more com-plicated by the additional constraint that the polymer must degrade in a predictable manner in terms of the rate and products of degradation. Our prejudice related to polymers for medical use is based on the complexity of such systems. Not only must we devise a structure that is conducive to cell growth and spreading, but cells must attach without distrupting their functions. During development of a colony of clin-ically useful size, the cells must be oxygenated and be provided with nutrients. Since vacularization will not have had time to develop, the void volume of the scaffold must fulfill that role by forming what we call pseudovasculature.

As the cell population grows it is hoped that blood vessels will develop. The scaffold must maintain sufficient integrity in order to prevent collapse. While these problems are widely recognized, clever researchers will eventually devise innovative technologies to solve them. We will discuss ideal scaffolds for artificial organs in the next chapter

Gunatillake and Adhikari[72] have written a very useful review of current biode-gradable chemistries. Their review focuses on implantable systems. It is important to keep in mind that the technologies have separated into two camps. The biode-gradable people focus on implantable devices and proponents of biodurability are working toward external devices. It is our opinion that both groups should merge around a single scaffold for the development of a hepatic colony and leave the location of the device to future researchers. The scaffold should at least be reticulated and preferably composed of polyurethane. This will become clear as we review the Gunatillake paper on chemistries and fabrication techniques for the current crop of so-called biodegradable scaffolds. The introduction defines tissue engineering as a "multidisciplinary field that involves the application of the principles and methods of engineering and life sciences towards the fundamental understanding of structure–function relationships in normal and pathological mammalian tissues...." The authors then review the basic chemistries currently considered for biodegradable scaffolds including polyglycolic acid (PGA), polylactic acids (PLAs), their copolymers (PLGA and others), and polycaprolactone (PCL). The degradation products of these polymers are present in the body and are removed by normal metabolic pathways. Certain natural polymers are also considered; collagen is the most common. While natural polymers, like collagen, are nominally acceptable, problems arise when they are processed. A number of the techniques are discussed below.

SOLVENT CASTING–PARTICULATE LEACHING

This technique involves producing a solution of PLA in a solvent. Salt particles are added. As the solvent evaporates, the salt particles become imbedded in the polymer mass. Soaking the polymer in water dissolves the salt and voids remain.

GAS FOAMING

The polymer is saturated with carbon dioxide (CO_2) under high pressure. As the pressure is released, the gas bubbles that form become the pores of the foam.

FIBER MESHES AND FIBER BONDING

Nonwoven scaffolds have been made from fibers of several biodegradable polymers.

FREEZE DRYING

Scaffolds are made by freeze drying a dispersion or solution of collagen. Freezing the dispersion or solution leads to the formation of ice crystals that force and aggregate the collagen molecules into the interstitial spaces.

The authors described several other fabrication techniques, but their conclusions are the important parts of their report: "Conventional scaffold fabrication techniques are incapable of precisely controlling pore size, pore geometry, spatial distribution of pores and construction of internal channels within the scaffold." They also state that scaffolds produced by the solvent casting–particulate leaching technique cannot guarantee interconnection of pores because interconnection is dependent on whether the adjacent salt particles are in contact. Moreover, only thin scaffold cross sections can be produced due to difficulty in removing salt particles deep in the matrix.

Only 10% to 30% of the pores were interconnected in the gas foaming process. Nonwoven fiber meshes exhibited poor mechanical integrity. Mass transport of nutrients, oxygen, and cells becomes a problem with materials that are not completely open. According to the observations of Gunatillake and Adhikari,[72] the following characteristics were identified as crucial properties of a scaffold:

- Pores of appropriate size to favor tissue integration and vascularization
- Biodegradability or bioresorbability
- Tendency toward cellular attachment, differentiation, and proliferation
- Adequate mechanical properties
- Inability to induce inflammatory responses
- Easy fabrication into a variety of shapes and sizes

Based on these requirements, the authors dismissed the current focus of research on implantable devices; the dismissal was premature, however.

PROPERTIES AND BIODEGRADATION OF POLYURETHANES

Earlier in this chapter, we suggested that a polyurethane could be made more durable by replacing the polyether polyol with a carbonate-based diol.[68] The object of this discussion is to illustrate how the degradation process can be encouraged. The philosophy is to develop a backbone that hydrolyzes in a controlled manner *in vivo*. Polyesters are usually used to accomplish this goal because esterase enzymes are present in the natural environment. Commercial polyester polyurethanes are useful without alteration but their degradation is considered too slow. The simplest method for making a so-called biodegradable polymer is to insert a molecule in a backbone for which there are natural enzymes.

Lactic acid is an example. It is conveniently added to the polyol and upon reaction with the isocyanate, it is inserted into the backbone. Upon exposure to fluids in the body, the polymer is cut at the insertion site. The reduction in tensile properties is dramatic. Although the fragments are suspect, the result is by definition degradable. A number of researchers at W.R. Grace[73] prepared biodegradable hydrophilic polyurethanes by a similar technique but the product was never available commercially.

Woodhouse[74] described a degradable polyurethane prepared with a novel amine chain extending agent. As with the insertion of lactic acid into the backbone, biodegradability was achieved, but the fragments appeared problematic. Thus, while biodegrading polyurethanes are fairly simple to prepare, one must be aware of the product to which a polyurethane degrades. The use of TDI as the isocyanate would clearly raise concerns from the FDA.

A more appropriate strategy may be to affect the chain cleavage within the isocyanate and then at multiple points within the polyol. One interesting approach is to use an amino acid-based isocyanate. While no such products are available commercially, Storey teaches the preparation of lycine diisocyanate.[75] In 1994, Storey prepared a number of polyurethanes with a lycine-based isocyanate using a number of degradable polyols including poly(dl-lactide). Several copolymer triols with acceptable elongations and tensile strengths were prepared. Degradation tests showed a large range of times, depending on the polyol used. In polyurethanes with high lactide contents, degradation time was as little as 3 days.

Bruin et al.[76] evaluated polyurethanes based on lycinediisocyanae (LDI) and poly(glycolide–caprolactone)diol for use as artificial skin. Implantation in guinea pigs showed rapid cell ingrowth and almost complete degradation in 4 to 8 weeks. No adverse tissue reactions were observed. They also reported the synthesis of a degradable polyurethane using so-called star-shaped polyester prepolymers prepared from a pentahydroxy sugar molecule by ring-opening copolymerization of l-lactide or glycolide with caprolactone. The prepolymers were cross-linked with 2,6-diisocyanatohexanoate. The degradation products of these PUR networks were considered nontoxic. The polymers were elastomeric; elongation ranged from 300% to 500% and tensile strengths varied from 8 to 40 MPa. Preliminary experiments in guinea pigs have shown that the polyurethanes biodegrade when implanted subcutaneously.

Zang et al.[77] developed a peptide-based polyurethane scaffold for tissue engineering. LDI was reacted with glycerol and upon reaction with water produced a porous sponge due to liberation of CO_2. Initial cell growth studies with rabbit bone marrow stromal cells showed that the polymer supported cell growth.

Hirt et al.[78] and De Groot et al.[79] reported the synthesis of degradable polyurethanes based on LDI, 2,2,4-triethylhexamethylene diisocyanate, and a number of polyesters, copolyesters, and polyethers. The polyurethanes ranged from elastomers with elongations at break as high as 780% (but low tensile strengths). Spaans et al.[80] reviewed microporous polyurethane amide and polyurethane urea scaffolds for repairing the knee joint meniscus. The soft segments in these polyurethanes were based on 50:50 l-lactide:PCL; chain extenders were adipic acid and water. Animal studies showed rapid cell ingrowth with no adverse tissue reactions. These further studies indicate that polyurethanes are the tools product designers can use to design efficient devices.

To close this chapter, however, we want to briefly discuss an emerging biotechnology technique: control of the adhesion of attachment-dependent cells. Certain cells within the body are said to be anchorage-dependent in that they must attach themselves to matrices in order to function and survive.

CELL ADHESION

Earlier in this chapter we discussed the designs of polyurethanes that resist protein adsorption. As a mechanism for developing a hemocompatible device, we noted the use of high molecular weight polyethylene glycol in an attempt to make the surface neutral with respect to the inflammation cascade. We covered the effects of hydrophobic surfaces and the development of fibrin capsules to isolate foreign bodies from their *in vivo* environments. We also discussed the use of polymers as scaffolds for cell growth and the adjustment of surface chemistry to provide a surface with the capacity to hold nutrients as a buffer for cell development. We also encouraged the development of bacterial colonies for processing contaminated fluids.

The final section of this chapter and the next chapter will cover a juxtaposition of these factors to effectively control the development of colonies of mammalian cells on the surfaces of scaffolds for medical remediation purposes. In the last chapter biofilters for environmental remediation were discussed. In a sense, such devices serve as biofilters for the body. In this regard, the liver, which support human cells, functions as a remediation device and as part of the endocrine system. An effective design would be a scaffold of high surface-to-volume ratio, high void volume, and significant physical strength that has a biocompatible internal surface and certain conformational properties that encourage the spreading of cells. These design features of an implanted device not surprisingly would also be components of an ideal biofilter for an environmental remediation device.

Animal cells add a significant layer of complexity to the system due to their dependence on cell–matrix and cell–cell signaling pathways. Unicellular species such as bacteria and yeasts tend to grow and proliferate as fast as nutrients can be supplied. In fact, their growth rates are typically proportional to the amounts of nutrients available. By contrast, however, cells from multicellular organisms must develop mechanisms that include both nutrient supplies and signal pathways that control cell division. Thus, while nutrients are necessary for an animal cell to proliferate, the cell must receive stimulatory chemical signals from other, usually adjacent, cells.

Certain cells of the body are said to be anchorage-dependent because they must attach themselves to matrices in order to function and survive. Without such attachments, they are subject to programmed death (apoptosis). Compounds known as integrins govern cell adhesion and also provide the signaling mechanisms necessary for cell survival. It is thought that this signaling involves the activation of growth factors.

One design requirement is to provide a physical scaffold that permits near-normal proliferation of cells. Proliferation is encouraged by scaffold with an open and curvilinear structure — a more or less natural conformation. For example, hepatic cells cultured on a flat plate do not function. The cell survival signals are functional as evidenced by our ability to keep cells alive, but their metabolic functions are destroyed when they are forced into a flat conformation. By inference, therefore, a concave structure is thought to be appropriate. As cells migrate to coat a surface, it is equally important to provide a scaffold that has as open an architecture as possible so as not to inhibit spreading.

The final and perhaps most difficult requirement is providing a surface to which cells can adhere. Studying the process of attachment is perhaps the newest area of biological science. An artificial scaffold is needed for cell attachment. Thus, it is important to include attachment as a design feature. This presents a challenging objective to polyurethane chemists, but the versatility of the chemistry will allow them to meet the challenge.

Anchorage dependency has been recognized as a requirement for cell viability[81] and the shape of a cell after adhesion is a determinant of cell expression. Hohner and Denker[82] examined hormone production from cells attached to substrata of both rigid and soft hydrogels of the same composition. The cells attached to the soft gel were more productive and this was attributed to the fact that the cells were more rounded. Mescher[83] studied cytotoxic T lymphocytes and latex particles of various diameters coated with ligands for T-cell receptors. An inverse relationship between

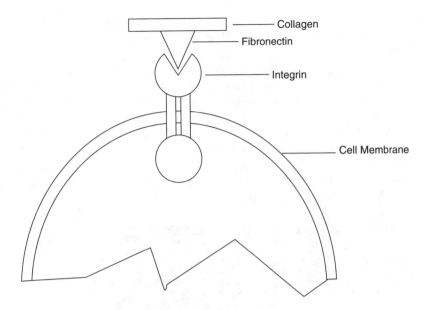

FIGURE 6.4 Attachment of cell to collagen by integrin.

cell size and ability to activate lymphocytes was observed. It is clear from these data that attachment is required, and also that attachment must be to a surface that does not affect cell–cell interactions. It seems appropriate to consider concave surfaces instead of beads.

Integrins, the proteins responsible for binding, govern both adhesion and the conformations of cells. They therefore indirectly control the cell–cell signaling pathways that determine proliferation and viability. Integrins are also responsible for the spreading of cells. Figure 6.4 depicts the attachment of a cell to collagen by integrin. The integrin is resident in a cell. It passes through the walls of cells anchored by actin filaments. Once a cell contacts a surface, it seeks to bind to it directly or indirectly through fibronectin (as shown in the figure).

Builders of artificial scaffolds are most concerned with this junction. In a natural environment, the binding is done via an intercellular glue known as extracellular matrix. The matrix is a protein gel produced by the cells and it serves as the basis of three-dimensional cell structure development. In the development of a three-dimensional scaffold, it is advisable that this mechanism be mimicked by some means.

In this regard, polyurethane technology offers a product designer a particular advantage. An aqueous solution or dispersion can be emulsified conveniently with a hydrophilic prepolymer and thus incorporated into the polyurethane matrix. The incorporation is accomplished by covalently bonding within the polyurethane backbone and by entrapping it within the matrix. Both methods are evident in foams produced by this technique.

The reactions of some protein with isocyanate removes some of the reactant from its normal role in the production of CO_2. This detracts from the quality of the foam. Because foam quality is critical to fulfilling the physical requirements of an efficient scaffold, this suggests that using hydrophilic chemistry in conjunction with a subscaffold of reticulated foam would be a more efficient use of this cell binding technique.

Our laboratory, in conjunction with the Maine Medical Center Research Institute, conducted a series of experiments to study the incorporation of cell binding components. A composite was produced by grafting an MDI-based hydrophilic polyurethane to a 30-pore-per-inch polyether reticulated foam using no surfactants. A fibronectant solution was added to discs of the composite foam. The discs were then inoculated with endothelial cells and cultured.

Microscopic evaluation of the pores of the foam showed that the cells took up residency along the structural members of the foam. Sampling on such a structure was problematic, but adhesion was clearly accomplished (Figure 6.5).

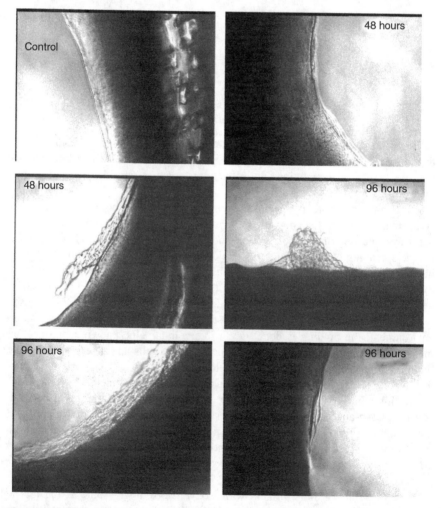

FIGURE 6.5 Cell adhesion on a hydrophilic polyurethane-grafted reticulated foam imbibed with fibronectin.

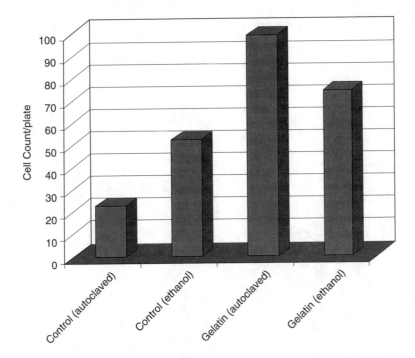

FIGURE 6.6 Adhesion of endothelial cells to hydrophilic polyurethane and gelatin copolymer composite (sterilized by ethanol or autoclaving).

It was hypothesized that the imbibed fibronectin may have been washed out by the medium, and this could have affected adhesion in sections of the disc. To eliminate this possibility, samples of the composite were made with gelatin dissolved in the aqueous phase. They were then emulsified with the hydrophilic prepolymer and applied to the reticulated foam. The result was covalent bonding of some of the protein into the polymer backbone, thus making it immune to washout. Figure 6.6 reports counts of endothelial cells cultured on this formulation and effects of sterilization method compared to controls without gelatin.

CONCLUSION

We have tried to show that polyurethane chemistry offers product designers the versatility to design unique polymer systems by balancing architecture and chemistry. A unique example is its use for cell adhesion and development of scaffolds.

7 Development of Artificial Organs

This chapter will review current activities and proposed research related to the development of artificial organs and organ-assist devices. While we will focus on the human liver, the discussion is applicable to other organs. The pancreas and the endocrine functions of the kidney follow the same basic path — the culturing of cells on a scaffold. The cells, of course, must function as they do in a natural uncompromised state while the scaffold provides a permanent or temporary template on which the cells attach and proliferate.

The mammalian liver is a construction of living cells that, uniquely among internal organs, simultaneously detoxifies, metabolizes, and synthesizes proteins. It allows the breakdown and synthesis of carbohydrates, lipids, amino acids, proteins, nucleic acids, and coenzymes. The liver is interconnected with other organs (pancreas, spleen, intestine) along the portal venous circulatory system. Clearly, an *ad hoc* view of the liver and its function to remove toxins, for example, is insufficient.

Fulminant liver failure results from massive necrosis of liver tissue with a concurrent buildup in toxic products, disruption of acid balance, and a decrease in cerebral blood flow. Impaired blood coagulation and intestinal bleeding result. The main physiological effect is diminution of mental function that often leads to coma.

Other malfunctions and diseases of the liver include viral infections and alcoholic hepatitis. In 1999, of the 14,707 individuals who were on the waiting list for liver transplants, 4,498 received transplants and 1,709 died.[83] As of February 2002, 18,434 people awaited liver transplants.

Whole or partial liver transplantation is considered the treatment of choice, but the need outstrips the supply. The development of a device that, at least temporarily, could serve the function of the liver would bridge the gap until a suitable donor became available. It could potentially reduce the load on a compromised liver until regeneration restored full function. It could also be helpful to patients who are ineligible for transplants.

In association with Dr. Len Trudell,[84] we began studying polyurethane as a scaffold for the propagation and use of human hepatic cells. The work related to what was then considered the most effective system for use as an extracorporeal artificial liver — the so-called hollow fiber technique. Our work addressed what we felt were significant problems: surface area and flux rate. It was felt that the chaotic structure of a polyurethane foam would lead to a greater effective surface. In addition, the hydrogel nature of the polyurethane we employed would produce high rates of transfer and broader and more carefully controlled molecular weight cutoff.

In early experiments with polysulfone fibers, we included a hydrophilic prepolymer (precursor for the polyurethane employed) in the polysulfone system. It was clear that while the prepolymer would have increased flux rate, it diminished the strength of the fibers enough to make them unusable. Although the use of the prepolymer in hollow fibers was not appropriate, the flux rates and cell structures of foams made from the polyurethane prepolymer were appropriate alternatives to hollow fibers if a construction could be achieved.

Trudell and our group, with the generous support of Dr. Hugo Jaurequi, conducted certain cell-culturing experiments on human hepatic cells. Since the effort was self-funded, the scope of the experiments was limited. Nevertheless, we felt we confirmed the viability of a specially designed polyurethane formulation as a substrate for cell growth.

This study was reported with another set of experiments that confirmed the growth of fibroblasts in connection with an experimental burn dressing.[84] While our work was positive enough to warrant further research, a technical problem prevented continued work. While we felt the surface chemistry was a necessary component of a successful artificial liver, it was not sufficient. In order to build a successful device, fluids must be able to enter and exit the environment freely. We discussed the differences in structure between open-celled polyurethanes and reticulated foam earlier.

Without a reticulated structure, it would not be possible under physiological conditions to oxygenate and feed the cells. For the same reason, blood or plasma would not be able to flow through a device to take full advantage of the high surface area. In order to proceed with the technique, this mass transport problem had to be solved.

Around that time, Taku Matsushita and coworkers began work on the use of conventional polyurethane foams (we used hydrophilic polyurethanes) as scaffolds for the propagation of hepatic cells. Using a rat model,[85] they were able to demonstrate the development of the hepatic spheroids necessary in the development of a functional artificial liver.

The success of Matsushita's method and the encouraging developments in our laborabory set the stage for a decade of independent research into what we see as the best technology for the development of a liver-assist device. In a sense, Matsushita's work confirmed the structure of the device and our work confirmed the chemistry of the surface. Matsushita continued his work without the benefit of our technology and has successfully demonstrated the use of his device in a dog model. In a 1999 report,[86] an artificial liver was reported to be "equal, and probably superior" to the most successful hollow-fiber device.

Cell transplantation has been proposed as an alternative to whole organ transplantation.[85,86] Much of the work conducted in the U.S. focused on implantable devices. Our work, the work of Matsushita et al., and the continuing work on hollow fibers focused on extracorporeal and paracorporeal devices.

This chapter will review the current state of the art in this arena. We first discuss the current methods for treating liver failure, including cell transplant techniques. The primary physical characteristics of successful implantable and extracorporeal devices will be described. We will then cover the work of Matsushita et al. that meets most of the requirements for a successful device along with improvements that will result from the use of hydrophilic-grafted polyurethanes.

The first section discusses current treatments for failed or failing livers including anticipated technologies that are under investigation in efforts to relieve or eliminate the shortages of transplanted organs. Among these techniques are transplanted cells, implantable tissue engineered devices, and extracorporeal systems. An excellent review of the current state of the art in the cell-based treatment of liver failure is given by Allen and Bhati.[89] In the next section, the properties of an ideal scaffold for use as a hepatic cell support system will be described with emphasis on biocompatibility and engineering aspects of both implantable and extracorporeal devices and significant projects in various stages of research and clinical trial. We will then demonstrate that a polyurethane or polyurethane composite could serve as an ideal scaffold for organ-assist devices.

CURRENT AND ANTICIPATED TECHNOLOGIES IN TREATMENT OF LIVER DISEASE

The treatment of liver disease has improved greatly with the development of transplantation techniques including surgical methods and suppression of the immune response. However, few liver diseases are curable, and surgical replacement is expensive and problematic from the perspectives of both the live donor and the recipient. Because the demand for transplantable livers far outstrips availability, alternative liver therapies are critical.

Several current technologies including cell transplantation, tissue engineered constructs, and extra- and paracorporeal devices seek to relieve some of the demands placed on a compromised liver and allow the liver to regenerate its function (liver-assist devices). Other devices seek to maintain patients until suitable donors are available (bridge-to-transplant devices). Several therapies seek to remove toxins from the blood (a function of the liver), and they too have places in treatment schemes. Based on the complex and multifunctional nature of the liver, however, some type of cell-based therapy is considered a more complete solution.

Advances in liver cell biology have provided valuable insights into the functioning of the organ. A flat plate of cells cannot function as an artificial liver. Liver cells are functional only when they can build three-dimensional (spheroid) structures. The next generation of devices must combine current understanding of the biology of hepatic cells with knowledge of cellular microstructures and the structures in which the cells will grow.

SURGICAL APPROACHES

Whole and partial liver transplants have become the treatments of choice for patients with imminent liver failure. Advances in surgical techniques and immune suppression are responsible for this progress. Treatments for liver diseases were unknown in the 1960s. Despite our improved knowledge of the function and physiology of the liver, one of every 10 individuals in the U.S. has been or will be diagnosed with liver disease. Hepatitis C virus (HCV) is a public health problem; approximately 170 million people are infected worldwide, and 8,000 to 10,000 deaths per year occur from HCV complications in the U.S. alone.[88] Few liver diseases are curable and the standard treatment is still transplantation. The following section surveys strategies for liver transplantation using recently deceased (cadaveric) donors, split-liver techniques, related living donors, and nonhuman sources.

The first human liver transplantations were performed in 1963 by Starzl and colleagues.[89] Until the application of immune system suppressants, the long-term survival rate of transplant recipients was poor. Continued improvements in surgical techniques, organ preservation, and immune suppression led to 1-year survival rates of 85% to 90%. The most common indications for the need for a liver transplantation are chronic hepatitis, alcoholic liver disease, and cirrhosis. The widening gap between the need for transplant material and the supply has caused the medical research community to search for alternatives to transplantation.

Researchers developed techniques to split a single liver from a cadaver and implant it into two recipients, thus increasing the supply of tissue. This measure originated as a means to reduce the sizes of transplant. The intent was to use the right lobe portion of a full-sized adult liver in a child and the left lobe in an adult. Split liver transplants approach the success rate of full transplants.[92]

Liver transplants from live donors began in the late 1980s.[93] A child can receive the left lobe of a liver from an adult with little risk to the donor.[94] Right lobe transplantation, however, presents a significant risk to the donor. Even with these developments, the supply situation demands that alternatives to transplantation be developed. An instructive series of slides can be accessed at livertransplant.org/patientguide/index.html. The slides walk a viewer through the transplant of a whole liver.

While liver transplantation represents a significant development for those with liver disease, a need continues to exist for therapies that address the risks of this major surgical procedure. Furthermore, the supply of tissue for transplantation will be a continuing problem. Alternative techniques are still experimental but in many cases appear promising. The next section introduces these techniques and provides an effective transition to the application of reticulated polyurethanes.

CELL-BASED APPROACHES

The answer to the shortage of liver tissue is to evolve from a dependence on whole and partial organs to the use of hepatic cells. Cell therapies range from the injection of cell colonies with the hope that they will take up residence and become clinically active to the development of implantable or extracorporeal devices. Such approaches must consider both the sources of hepatocytes and stabilization of liver-specific functions.

CELL SOURCING

The selection of cell types for these therapies is an important part of device design. Because these therapies represent emerging technologies, our understanding of individual and aggregate cell functions is less than complete. Most current technologies in this field use primary hepatocytes. Many devices in clinical study are based on the use of pig hepatocytes and present peculiar engineering challenges based on obvious incompatibilities. Porcine cells are less well characterized than rodent cells.[93]

In direct cell implantation, human cells are required but supplies are limited. Production of hepatocytes from pediatric patients has been reported and techniques for preservation are improving.[94,95] *In vitro* characterization of human hepatocytes will aid the development of improved cell-based therapies. The primary safety concern with the use of cell lines, especially implanted cells, is the transmission of toxic factors to the host.

Several sources have reported using stem cells including embryonic stem cells, adult liver progenitors, and transdifferentiated nonhepatic cells in cell-based therapies.[96,97] Hepatocyte lineage *in vitro* has been reported in murine embryonic stem cells.[98] It is apparent that hematopoietic stem cells can generate hepatocytes directly. This has been shown in rodent models and confirmed in humans by a study of recipients of bone marrow and liver transplants.[98]

The success of cellular therapies ultimately depends on the stability of the hepatocyte in the architecture in which it must exist. Primary hepatocytes are anchorage dependent. Isolated cells rapidly lose viability when cultured in monolayers or suspensions. Investigators have developed culture models based on features of liver architecture to recapitulate the complex hepatocyte microenvironment. Sandwich culture mimics the environment of hepatocytes *in vivo* by entrapping cells between two layers of collagen gel.[99] However, such methods introduce additional transport barriers and are difficult to scale up to therapeutic levels.[100]

Cell–cell interactions and cell–scaffold interactions have been shown to improve viability and stabilize function.[101] Complete understanding of the mechanisms required to stabilize hepatocyte function will have a broad impact on this technology.

CELL TRANSPLANTATION

It is possible to produce a suspension of hepatic cells and implant the cells by injection. Cells implanted in the liver and even the spleen become viable.[102,103] The procedure is not without problems, however. The yield of viable cells is questionable. A critical element for effective regeneration is the availability of sites for cell growth. The implanted cells also need significant time for engraftment and proliferation (doubling time in mice = 28 hours).[104]

TISSUE-ENGINEERED IMPLANTS

Implantable tissue-engineered constructs are derivatives of cell therapies. They have certain advantages over implanting cells in that the culturing process is more easily controlled and tested before risking implantation. The technique is highly experimental and not without problems. We cover this more completely in later sections.

Despite advances in hepatocyte culture, tissue engineering of the liver faces significant challenges.[105] One important factor in the development of a self-sufficient hepatic device is vascularization. While in the early stages of development, a device can be maintained by providing sufficient void volume; the device ultimately will have to provide for natural supplies of raw materials. This is one of the factors that makes the development of a biodegradable implanted device problematic. Not only must the so-called biodegradable scaffold (and its associated problems) be resorbed or otherwise eliminated, it must do so within the time frame required to build vasculature. This detracts from the practicality of a biodegradable system and is a major reason for our focus on biodurable devices.

EXTRACORPOREAL DEVICES

Extracorporeal devices to support a compromised liver were reviewed by Allen et al.[89] and Strain and Neuberger.[106] Various nonbiological approaches such as hemodialysis or hemoperfusion over charcoal have met with limited success, presumably because these systems inadequately replaced the synthetic and metabolic functions of the liver.[107] Conversely, biological approaches such as hollow fiber devices, flat plate systems, perfusion beds, and suspension reactors have shown encouraging results but are difficult to implement in a clinical setting.

The most common bioartificial liver device design incorporates hepatocytes in hollow fiber cartridges borrowed from hemodialysis. Hollow fiber membranes provide scaffolds for cell attachment and immuno-isolation, and are well characterized in clinical settings. Flat plate or monolayer bioreactors offer better control of the hepatocyte microenvironment and minimize transport barriers. However, it may be difficult to provide uniform perfusion of the packing matrix, and cells can be exposed to damaging shear forces.[108] A successful extracorporeal bioartificial liver design will include sufficient mass transport to supply oxygen and nutrients and allow for sufficient cell population growth. As noted earlier, hepatocytes require a specific microenvironment to maintain liver-specific functions.

The most advanced technology is the extracorporeal hollow fiber reactor. It is currently in Phase III trial and achieved a good Phase II record to support it. Other techniques including a polyurethane system devised in Japan[97] and encapsulated hepatocytes from UCLA[109] are or were in large animal trials. Whether a device is extracorporeal or is intended for implantation, clinical significance requires a suitable scaffold to support a sufficiently large colony of hepatic cells. For both extracorporeal and implant use, the physical structure of the scaffold must meet certain requirements of strength, void volume, biocompatibility, and other parameters.

DESIGN OF IDEAL SCAFFOLD FOR EXTRACORPOREAL BIOARTIFICIAL LIVER (BAL) OR IMPLANTABLE ARTIFICIAL ORGAN

Yang et al.[110] discussed the basic properties of an implantable or extracorporeal artificial liver. The article focused on implantable devices but other than biodegradability, the properties of implantable devices are also applicable to extracorporeal devices. The focus of the article on implantable devices reveals an unfortunate prejudice on the part of much of the scientific community. Most researchers in this field are working on devices intended to be placed in the body.

Yang's paper implies that the ideal artificial organ should be permanently implanted in the body. It is our feeling that this restriction makes the program needlessly complicated. Our philosophy is that a program to design an intermediate device to assist a diseased or compromised liver is an easier target and has the potential to save more lives in the short term. Once a viable autologous liver-assist device becomes a reality, we will learn from it and then be able to concentrate on the much more difficult implantable devices.

The knowledge gained from this novel cell transplantation device will allow us to expand the technology to other organs and systems, for example, an artificial pancreas, kidney, or extracorporeal bone marrow system. Nevertheless, Yuang offers a glimpse of what would be required of a scaffold for use in a cell transplant device.

Tissue engineering, an important emerging topic in biomedical engineering, has shown promise in creating clinically useful devices from harvested tissues.[111] The underlying concept is that cells can be isolated from a patient and cultured *in situ* in or on a scaffold carrier. The resulting tissue engineered construct is then grafted back into the patient or built into an extracorporeal device to function as a replacement or assist a compromised organ. Both approaches require highly porous artificial extracellular matrices[112] or scaffolds to accommodate mammalian cells and guide their growth and tissue regeneration in three dimensions.

Three-dimensional scaffolds for tissue engineering have been found less than ideal for a number of complex reasons ranging from mechanical strength to cell structure to hemocompatibility. Several requirements impact the designs of scaffolds for tissue engineering. In addition to biocompatibility and hemocompatibility, scaffolds should possess appropriate mechanical properties to provide the correct stress environment. A scaffold should be porous and permeable to permit the ingress of cells and nutrients, and it should also exhibit the appropriate surface structure and chemistry for cell attachment. It should have a high surface-to-volume ratio and a surface shape conducive to the three-dimensional growth of cell agglomerates. The material should also have a natural or synthetic ability to bind to cells using some natural cell adhesion process. A scaffold should have sufficient void volume to provide for nutrients and oxygen for the developing cell structure, not to mention acceptable blood characteristics of flow rate and pressure drop (commonly called the mass transport phenomenon). Finally, a scaffold should be able to accommodate a membrane to separate blood flow from nonautologous and xenographic hepatocytes. In summary, the ideal scaffold has the following properties:

- Biocompatibility and hemocompatibility
- Appropriate mechanical properties
- Appropriate structural aspects: void volume, pore size, interconnected pores
- Provision for cell attachment
- High surface-to-volume ratio
- Ability to covalently attach ligands
- Physical environment for cell proliferation and spreading

BIOCOMPATIBILITY AND HEMOCOMPATIBILITY

The first issue with regard to tissue engineering is the choice of suitable material. The desirable characteristics of these materials are biocompatibility (not provoking an unwanted tissue response to an implant and possessing the appropriate surface chemistry to promote cell attachment and function). Potential materials with these characteristics include natural polymers, synthetic polymers, ceramics, metals, and combinations of these materials. Several semisynthetic polymers have been investigated as possible candidates for use in biodegradable scaffold systems. These polymer systems were discussed in Chapter 6.

In an ideal case, a liver-assist device would be compatible with whole blood. The goal is a device that can directly enter the bloodstream (by implantation or as an extracorporeal device). This task is not an easy one. The hollow fiber technologies illustrate the paradox. In order to ensure that hollow fibers have sufficient strength, they must be made from a somewhat hydrophobic synthetic material. Two problems arise as a result of this compromise. First, the flux rate is decreased, which in turn requires that the system operate at higher than physiological pressures. This calls for higher-strength materials that reduce flux rates and lead to larger devices that can handle the higher pressures. The compromises are in areas that are not beneficial to the overall effectiveness of the device. Second, because the materials used for the device are hydrophobic, they initiate inflammatory responses in the blood. For this reason, an ideal scaffold would be of a material compatible with whole blood. Dixit et al.[109] showed this to be an achievable goal.

STRENGTH OF MATERIAL

A scaffold should have the mechanical strength needed for creation of a macroporous device that will retain its structure. The biostabilities of many implants depend on factors such as strength, elasticity, absorption at the material interface, and chemical degradation. Processability of the biomaterial is also required when the final shape of the repaired organ or regenerated tissue has a critical influence on its activity. A scaffold should be easily processed into a variety of configurations. The reproducibility of scaffold or architecture is also vital to maintaining dimensional stability.

Scaffolds can also be formulated to contain additives or active agents for more rapid tissue growth or compatibility. These additives will doubtless have an effect on strength. A device in contact with blood and bodily fluids must be strong enough to withstand hydrolysis and enzymatic attack. It is important to note that the physical properties must be measured in the environment in which the device is to exist. Most of the materials considered for this are hydrophilic. For example, the strength of a material swollen in plasma should be the standard for measurement. With biodegradable devices, the rate of decrease in the strength of the material must be juxtaposed against the development of cell colonies and the rate of vascularization.

PORE SIZE AND STRUCTURE

The regeneration of specific tissues aided by synthetic materials has been shown to be dependent on the porosity and pore size of the supporting three-dimensional structure. A large surface area favors cell attachment and growth, whereas a large pore volume is required to accommodate and subsequently deliver a cell mass sufficient for tissue repair. Highly porous biomaterials are also desirable for the easy diffusion of nutrients to and waste products from the implant and for vascularization — major requirements for the regeneration of highly metabolic organs such as the liver and pancreas. The surface area:volume ratio of a porous material depends on the density and average diameters of the pores. Nevertheless, the diameters of cells in suspension dictate minimum pore size, which varies from one cell type to another. Depending on the envisioned application, pore size must be carefully controlled.

SURFACE-TO-VOLUME RATIO

Hepatic cells fall into a class known as anchorage-dependent cells. Therefore, colonization is in part a function of the surface area of the device. Clearly, a scaffold that provides the highest surface area in the smallest volume (offering the highest net density in cells) would be preferred.

MASS TRANSPORT THROUGH DEVICE

The device will have to hold within its volume a colony of hepatic cells and perhaps a hydrogel membrane and still have enough void volume to allow physiologically correct blood flow. Hepatic diameters spheroids are about 100 μm.[115] In order for a scaffold to contain a significant number of hepatic cells and still be able to handle the flow rates required, it must have a large void volume.

HIGH DEGREE OF INTERCONNECTED CELLS

One persistent problem in cell culturing is achieving high density and uniform coating. It is difficult to be quantitative about this aspect of design, but it is clear that the free movement of hepatic cells through the scaffold prior to adhesion would be beneficial in meeting the goal of uniformity. Figure 7.1 compares a candidate scaffold used by Funatsu and Ijima[113] with a poly(glycolic-lactic) acid scaffold.[114]

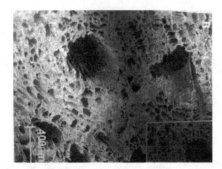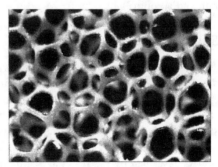

FIGURE 7.1 Visual comparison of cell interconnectedness of two candidate scaffolds. Left: the Hasirci[114] device. Right: A device similar to the one devised by Funatsu and Ijima.[113] Hasirci, V. et al. *Tissue Eng.*, 7:4, 2001. With permission.

Void Volume

While the scaffold is the structure upon which the hepatic cells must congregate, it must be constituted from a minimum amount of material. The device must be able to contain the hepatic spheroids that are in a constant state of flux and allow for blood to pass through in physiological conditions without channeling or other diminution of effective surface area. We will describe the concept of pseudovasculature that depends on a scaffold with a very large void volume. The use of pseudovasculature is intended as a temporary measure until natural vasculature develops.

Zeltinger[115] et al. studied the effect of pore size and void volume and their effects on cell adhesion, cell proliferation, and the deposition of cells in artificial scaffolds. Using a fabrication method known as Theriform, a multilayered porous structure was produced. While not truly three dimensional, the method nevertheless allowed the team to study the void volume (also known as void fraction) and calculate average pore size. With regard to pore size, the standard deviation was a high fraction of the average. In our opinion, this reduced the value of the conclusions of the paper. The void volume, however, is of great value. The ideal scaffold, however, would have a high void volume and a very narrow distribution of cell size. This is possible with a polyurethane reticulated foam. Despite its limitations, the Zeltinger paper concluded that "the optimum void volume for engineering homogenous tissues may be greater than 90%." As we will report, the volume for reticulated foams is approximately 94%. Other researchers studied the effects of void volume and pore size of scaffold.[116,117]

Allowance for High Flux Membrane

While the goal of our program will be the use of autologous cells in an assist device, it is clear that the road to that goal will begin with work on xenographic cells. Indeed, the major clinical studies use porcine cells. Therefore, it is a pragmatic requirement that a scaffold is capable of being fitted with a membrane that isolates the hepatocytes from blood. The membrane would, however, have to have a significantly high flux rate and a controllable cutoff.

Shape of Colonizing Surface

The emerging science of cell adhesion will doubtless play an important role in the development of an ideal scaffold. It is interesting to note that when the current best extracorporeal technologies and implantable devices were first conceived, our understanding of how and why cells bind to a surface was crude at best. While our current state of knowledge is still rudimentary, we know enough to be able to include it in our design set.

As described by Pierres et al.,[118] the process of adhesion occurs in three stages. The first stage (adhesion) consists of antibody–antigen coupling. The antibody is a transmembrane species and its complement antigen is in the surface. This is a rate determining step that can take as long as 45 minutes.[119] The initial stage is therefore a chemical interaction and will be covered in a subsequent section.

The second stage of binding is called fitting. The cell deforms somewhat by increasing the number of binding sites. While still using the antibody–antigen mechanism, it might also respond to van der Waals forces. The result is a conformation of the cell and the wall to which it is attached.

The third stage of binding is spreading. This involves cell reorganization and is dependent on metabolic inhibitors. The results are dramatic changes in cell structure and functional properties.

The binding process may exert a profound effect on cell function. It is well known that hepatic cells do not function when cultured on a flat plate. This reason may be partially explained by the binding–spreading phenomenon. It therefore follows that an ideal scaffold would minimize the spreading effect. Inasmuch as hepatic cells are our primary interest, a scaffold that best approaches the native shapes of the cells would appear to be the best. It is equally clear that this shape must be optimized against additional requirements of void volume and mass transport. Nevertheless, an ideal scaffold would be more curvilinear than flat.

ATTACHMENT OF LIGANDS

Cell adhesion and proliferation are said to be controlled by the surface chemistry of the scaffold. Cell adhesion is covered in the next section. Here we will describe the ability of the scaffold material to be derivatized with ligands. The materials commonly considered for this application can be derivatized with varying amounts of effort. Since many of the materials are already difficult to process, adding the complication of an attached ligand is very problematic. For process reasons, it would seem advisable that the ligand attachment be conducted after the scaffold is constructed. By way of example, hollow fibers are extruded through a die and therefore the melt or solvent flow is an important determinant of quality. Derivatizing the polymer before processing would change the melt and solvent flow characteristics. Given the types of ligands that would be used, this would be difficult.

CELL ADHESION

As stated above, the adhesion process begins with a chemical (not necessarily covalent) bonding followed by the adjustment of the cell shape to the substrate. In the design of an ideal scaffold, therefore, the two aspects that must be considered are the chemistry of the attachment and the conformation of the scaffold.

The chemical part of the adhesion process dictates that in designing a surface for attachment of cells, one must seek to stimulate an active interaction between the surface and the scaffold. The surface properties of the scaffold are our main concerns. The surface should mimic the natural support structures on the human body. Extracellular matrix (ECM) provides cells with an interactive structure onto which they can adhere. This process (referred to as integrin-mediated binding) is a basis of cell growth.

Much of the plethora of new research in cell-scaffold interactions should be considered in the development of the ideal scaffold. The implementation of these new technologies will depend on the incorporation and preferably covalent binding of the receptor and the maintenance of its activity during the culturing process.[120]

CURRENT CLINICAL ACTIVITY IN SCAFFOLD-BASED ARTIFICIAL LIVER DEVELOPMENT

By far, the most advanced technology in current use is the hollow fiber technique. It has been reviewed extensively in the literature. Briefly, this configuration involves the cultivation of hepatocytes on the external surfaces of semipermeable capillary hollow fiber membranes bundled together within a plastic shell. Nutrients and ultimately plasma from patient blood are circulated through the fibers. The cells in the capillaries provide hepatic function. In the current versions of this technology, cultured porcine hepatocytes are protected from the body's immune system by the semipermeable capillary membrane.

The most promising clinical results using this technique are from the work of Sussman's group[121] with the C3a human hepatoma cell line and the work of Demetriou et al.[93] with primary porcine hepatocytes. Each group used its own bioreactor design. In Demetriou's case, after excellent results in Phase II trials, the HepatAssist, as the device is known, is currently in Phase III testing in a multicenter study. Figure 7.2 shows how the device is inserted into the bloodstream.

Several other projects are showing some progress in clinical trials. See Table 7.1. The Vitagen project[122] using immortalized hepatic cells in a hollow fiber capsule in combination with a carbon-based toxin removal system is worthy of special note. The ELAD™ bioartificial liver support system is designed to combine the unique

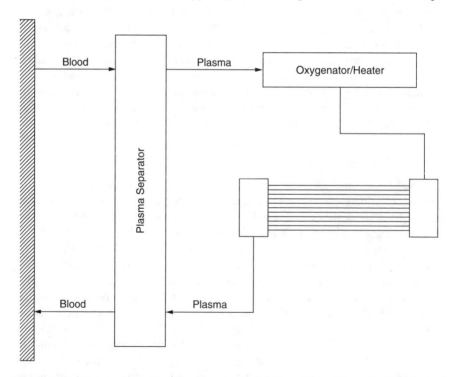

FIGURE 7.2 HepatAssist hollow fiber-based artificial liver (Circe Biomedical, Lexington, MA).

TABLE 7.1
Current Clinical Studies on Extracorporeal Devices

Company/Institution	Number of Patients	Clinical Phase	Comments
Vitagen[122]	25	I and II	Porcine cell
Circe[93]	171	II and III	Porcine cell
Excorp[123]	5	I	Porcine cell
Charite/Virchow Medical School[124]	8	I and II	Porcine cell

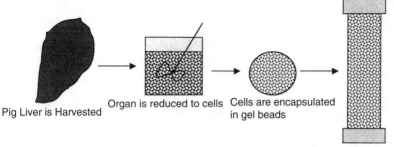

Pig Liver is Harvested   Organ is reduced to cells   Cells are encapsulated
                                                                 in gel beads

                                                             Gel beads are packed into a
                                                             flow-through column

FIGURE 7.3 UCLA bioartificial liver preparation scheme.

and potentially powerful benefits of cell therapy with a reliable medical device similar to that used in kidney dialysis treatment. As a cell therapeutic, the ELAD is believed to perform many essential biological liver functions including metabolizing toxins and producing beneficial proteins. Vitagen's proprietary C3A immortalized hepatic cells allow the ELAD system to provide extended continuous treatment (more than 7 days in recent clinical studies). Table 7.1 shows the current human trials of extracorporeal hollow fiber devices.

While these devices represent the leading edge in extracorporeal liver-assist devices, two other projects currently in animal trials deserve some attention as they represent other scaffold types.

The so-called UCLA bioartificial liver involves the direct hemoperfusion of microencapsulated porcine hepatocytes in an extracorporeal chamber (Figure 7.3). Since it permits perfusion with whole blood, it has an advantage over the hollow fiber technique that has to be perfused with plasma. The hepatocytes are isolated from pig livers and microencapsulated in an alginate–polylysine membrane. Microencapsulated hepatocytes are approximately 300 to 700 µm in diameter.

One of the philosophical bases of this work is the relative surface area benefit of microcapsules vs. hollow fibers. It is also felt that the alginate hydrogel would have a higher flux rate than the polysulfone used in hollow fibers. The study of this technique, however, has only been advanced through a rat model. The survival rate for induced liver failure was 46.7% vs. 0% for untreated animals.

TABLE 7.2
Comparison of Kyushu University's Polyurethane Foam Bioartificial Liver and Demetriou's Hollow Fiber Liver

	Demetriou (Hollow Fiber)		Kyushu University (Polyurethane Foam)	
	Baseline	6 Hours	Baseline	6 Hours
Blood pressure (mmHg)				
Control	187	50	137	75
Test	212	100	156	129
Glucose (mg/dl)				
Control	102	4	207	15
Test	142	48	184	124
Ammonia (mg/deciliters)				
Control	95	220	173	602
Test	89	186	123	235
Lactate (mg/dl)				
Control	21	116	10	81
Test	17	68	12	50

A group at the Department of Chemical Engineering at Kyushu University in Fukuoka, Japan, spent the last decade working on a polyurethane foam scaffold for hepatic cells.[125] In 1999, the group reported their work using a dog model that, showed "that the performance of this system was equal, or probably superior to that of Demetriou's system, and in addition, our system improved renal function." This opinion is based on the comparison of various blood chemistries in the two devices (hollow fiber vs. polyurethane foam). Table 7.2 depicts the comparison.

The basis for the Kyushu University technology is the culturing of hepatic spheroids within the matrix of a polyurethane foam. Figure 7.4 is a micrograph of a hepatic spheroid in foam. The above data were gathered during preclinical trials in a dog model. Figure 7.5 shows the circuitry of the study. The technology still requires that the plasma be separated from the blood. There are also issues of cell attachment, proliferation, cell spreading, and flow restrictions due to attachment problems.

Therefore, we could say that the reticulated scaffold chosen was appropriate based on its open pore structure and physical strength, but as we have seen in other applications, its chemistry is less than ideal. With this conclusion and our earlier work with Trudell that indicated that our chemistry was correct, we developed a process that combined the two aspects into a composite. We will discuss this in detail later in this chapter.

It is useful to examine reticulated foams in the context of their suitability as scaffolds for artificial organ development. Reticulated foams are made in a postprocessing step after polyurethane foams are made. Polyether and polyester polyurethanes are made for reticulation. The major use of the materials is in air filtration where resistance to flow (a mass transport property) is an important requirement.

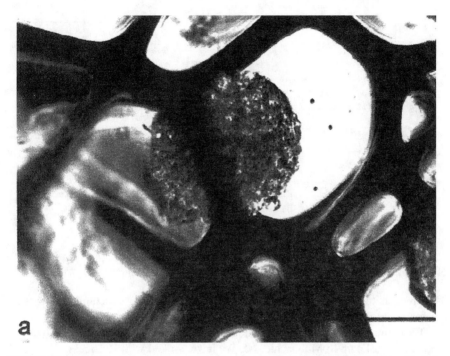

FIGURE 7.4 Hepatic spheroid in structure of polyurethane foam. (From Yamashita, Y. et al., High Metabolic Function of Primary Human and Porcine Hepatocytes, *Cell Transplantation,* Vol. 11, pp. 379–384, 2002.)

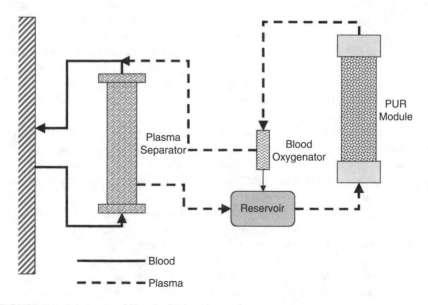

FIGURE 7.5 Schematic of Kyushu University study.

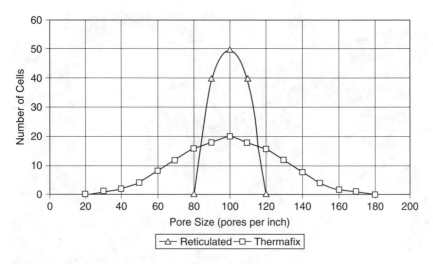

FIGURE 7.6 Comparison of Zeltringer and typical reticulated foam data.

The object is to remove residual materials between the major structural members of the foam and thus reduce the resistance to fluids flowing through it.

The dominant reticulated foam manufacturers in the U.S. are Foamex in Eddystone, PA, and Crest Foam in Moonachie, NJ. We have worked with Recticel in Brussels, Belgium, and find the company to be competent and cooperative. Our work has not involved companies in the Far East, but Japan probably has competent foam manufacturers. Both Foamex and Recticel are active in Eastern markets.

We discussed reticulated foams earlier in this book. They appear to have many desirable properties of ideal scaffolds. Depending on the feedstock, the manufacturers can produce a wide variety of pore sizes. Foams made specifically for reticulation have very narrow pore size distributions. If we compare the reported cell size distribution with that of Zeltringer, we can illustrate the precision of the reticulated foam process in the context of scaffolds for cell growth. Caution is advised in reviewing the Figure 7.6 plot. It is qualitative and assumes a normal distribution for both systems. It estimates the Zeltringer data based on the published standard deviation.

To provide hepatic cells sufficient room to propagate without materially affecting mass transport, it is important that a composite have a high void volume. Here too a reticulated scaffold offers certain advantages. The absolute density of a polyurethane is approximately that of water, while the foams we have been describing have densities around 2 lb/ft³. This mathematically leads to a void volume of 90% or more. One goal of the research project we propose is the study of this remarkably high void as a pseudovasculature, that is, the hepatic cells, after taking up residence within the scaffold, leave enough room for blood to flow.

Matsushita and colleagues saw the reticulated material as a viable scaffold for their research. We, however, saw it as the starting point for a scaffold. We were convinced that no single polymer phase was capable of reaching a viable optimum

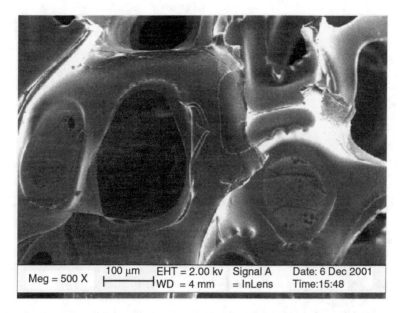

FIGURE 7.7 Micrograph of CoFoam® HPUR showing skeletal core (oval contrasted area at lower right).

between the ideal physical state of matter and the ideal chemistry for cell propagation and function. It was our strategy therefore to use the reticulated scaffold and insulate its hydrophobic surface by grafting a hydrophilic polyurethane polymer to it.[14] Figure 7.7 shows the product of this grafted coating.

The process of manufacturing this composite is described in Chapter 2, but in the context of its use as a scaffold, it was necessary that we maintain the physical properties of the reticulated foam in terms of void volume, surface area, etc. We pumped water through a capsule containing CoFoam HPUR. The study was conducted for another project, so flow rates shown in Figure 7.8 are not physiological. Nevertheless, the data testify to the suitability of CoFoam for the application proposed.

The last topic in evaluating the suitability of reticulated foam as the scaffold of a composite is somewhat qualitative. It is known that hepatic cells do not function when cultured on a flat plate.[87] At least part of the reason for this is the deformation of the cells developed during the spreading process.[126] It seems likely that if a cell is sufficiently deactivated by a flat surface, the effect will be as severe as when culturing on a convex surface such as the outside shape of a hollow fiber. A reticulated foam, however, presents a cell with several opportunities for a more natural attachment. The dodecahedron structure of each foam cell would appear to be a more natural scaffold for attachment. This perhaps explains the claimed superiority of a scaffold based on a reticulated foam of Gion et al.[125] over the HepatAssist hollow fiber device. Part of the research program that we will propose is that the effects of conformational aspects of an efficient scaffold will be quantified.

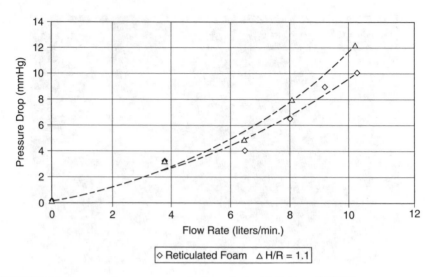

FIGURE 7.8 Pressure drop across a capsule packed with CoFoam® HPUR.

As we described in Chapter 6, a possible advantage of the system is the ease of attachment of ligands to the hydrophilic polyurethane. Copolymerization with attachment ligands or antigens is also conveniently accomplished. Also, the reservoir capacity of the hydrophilic polyurethane is useful for storing nutrients and acting as a buffer as noted in the discussion of biofilters in Chapter 5.

SUMMARY

We presented a case for considering polyurethane reticulated foam as a suitable scaffold for a liver-assist composite. This suggestion is far from conjecture as we showed. Researchers in Japan used a component of the scaffold we propose with some success. We reviewed their work to show that the physical properties of an ideal scaffold were fulfilled by their construction. We feel that a hydrophilic polymer grafted to a reticulate could produce an improved liver-assist device that could provide the required chemical functioning. As we established in our earlier work, the hydrophilic biocompatibility and hemocompatibility and provisions for cell attachment ligands are critical components of the ideal scaffold. The chemical flexibility including ligand attachment and the proper choice of polyols offers more advantages.

On March 1, 2002, the FDA granted special orphan drug status to bioartificial liver systems designed to support and enhance the ability of a human liver to regenerate. Under the Orphan Drug Act, the FDA can expedite approval and grant special status to companies that make drugs or devices for conditions that affect fewer than 200,000 people each year. The FDA approval should inject new vigor into the development of solutions to this important health problem.

8 Other Applications

This chapter covers a number of applications, none of which is sufficiently developed to justify a whole chapter. This is not a reflection on the importance, interest, or possible impact of the technology, especially the first section in which we discuss the use of polyurethane to immobilize enzymes and cells.

The architecture and versatility of reticulated foams offer engineering quality substrata on which enzymes can be covalently bonded. They confirm the thesis of this book: a practical solution is a combination of the correct scaffold and the correct chemistry. Immobilized enzymes could, by virtue of their capabilities, be awarded chapter status; our work in this area is still embryonic. Nevertheless, we will discuss the importance of this technology and show why polyurethanes may represent the breakthrough this area of science needs. We will focus as always on ease of immobilization (time and cost) and the flow-through architecture discussed earlier.

This chapter presents a number of consumer applications. While they do not garner the attention accorded to biomedical applications, they are challenging and commercially important. We will describe a hydrophilic polyurethane matrix as a suitable substitute for the common technique of emulsifying active ingredients in creams and lotions. A dozen or so ingredients may be used to create a stabile delivery form for active ingredients. Consumers appear desirous of eliminating petrochemical-based emulsifiers, stabilizers, and surfactants. We will show how this can be accomplished by embedding an active ingredient within a hydrophilic matrix. The active ingredient is then delivered to the surface of the skin in a "pure" form.

We will describe the use of polyurethane as artificial lobster bait. With the assistance of the Lobster Institute, we were able to trap about 55% as many lobsters with artificial bait as were trapped with natural bait (herring). Development continues; our goal is to achieve the 75% catch rate compared to herring.

IMMOBILIZATION OF ENZYMES AND CELLS

The concept of immobilizing enzymes and cells on a polymer scaffold has been studied for several decades in one form or another. The work was driven by the need to increase the half-lives of enzymes, which are notoriously short. Part of the problem is that enzymes at efficient concentrations degrade each other. Denaturation and poisoning are similarly detrimental to the life of an enzyme. It was hypothesized that some of these problems could be mitigated by attaching the enzyme to a substrate. Experience has shown this is an effective way to improve the efficiency of what can be very expensive biological catalysts.

It is not the purpose of this essay to describe how enzymes work. Many excellent texts cover the subject. Enzymes are components of the system by which living cells, plants, and animals produce energy and nutrients. Possible food sources in the

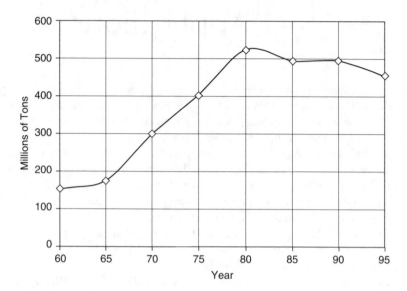

FIGURE 8.1 Pesticide usage for corn, soybeans, cotton, wheat, sorghum, potatoes, barley, rice, tobacco, sugar beets, peanuts, and oats, 1960 through 1995.

environment are rarely in forms that can pass through cell membranes to supply energy and materials for cell growth. Thus, enzymes are produced and emitted from cells (extracellular) or are produced and used inside cells to complete the digestion process.

We have learned to take advantage of the enzyme technology in the medical and chemical processing industries. Laundry detergents contain enzymes derived from high temperature-resistant bacteria living near volcanic vents. Several classes of enzymes are used to convert corn syrup into fructose to sweeten soft drinks. Enzymes are beginning to be applied to environmental problems, particularly those involving recalcitrant pollutants.

Pesticides are vital agricultural tools that protect food and fiber plants from damage by insects, weeds, diseases, nematodes, and rodents. U.S. agriculture spends about $8 billion annually on pesticides, representing about 70% of domestic pesticide sales. The dependence of agriculture on chemical pesticides developed over the last 60 years as the agricultural sector shifted from labor-intensive production methods to more capital- and chemical-intensive methods.

Use of conventional pesticides on farms increased from about 400 million pounds (active ingredient) in the 1960s to over 800 million pounds in the late 1970s and early 1980s, primarily due to the widespread use of herbicides in corn production. Since then, annual usage ranges from 700 million to 780 million pounds. Agricultural use can vary considerably from year to year depending on weather, pest outbreaks, acreage sown, and economic factors such as pesticide and crop prices. See Figure 8.1. While the quantity of pesticides used for agriculture has dropped slightly in recent years, total expenditures on pesticides by farmers are still increasing.[127]

TABLE 8.1
U.S. Organophosphate Usage

Use	Crop	Quantity Used (million pounds)
Agricultural	Field corn	19
	Cotton	15
	Other field crops	10
	Fruits and nuts	9
	Vegetables	7
Nonagricultural	Livestock and pets	4
	Residential	7
	Mosquito control	3
	Grain storage	2
	Turfs and ornamental plants	1

During the 1960s, insecticides dominated agricultural pesticide use, accounting for about half of all pesticides used. The quantity of insecticides fell as the organochlorines (DDT, aldrin, and toxaphene) were replaced by pyrethroids and other chemicals that required lower application rates. Today, pesticide levels in water are monitored routinely. Pesticide residues have been found in groundwater, surface water, and rainfall. EPA began to emphasize groundwater monitoring for pesticides in 1979 after discoveries of DBCP and aldicarb in groundwater in several states. In 1985, 38 states reported that agricultural activity was a known or suspected source of groundwater contamination within their borders.

Most of the common pesticides used now and in the past have also been found in the atmosphere, including DDT, toxaphene, dieldrin, heptachlor, organophosphorous insecticides, triazine herbicides, alachlor and metolachlor. These airborne pesticides return to the earth with rainfall to further contribute to water contamination. A recent U.S. Geological Service report of a survey of pesticides in the nation's waters concluded that pesticides were common in surface and shallow groundwaters in both urban and agricultural areas, but investigators were not able to determine whether contamination is decreasing or increasing.

About 60 million pounds of organophosphates (OPs) are applied to about 60 million acres of U.S. crops annually. Nonagricultural uses account for about 17 million pounds per year.[131]

The extensive use of this class of pesticides made it inevitable that significant quantities would find their ways into water supply systems. It is hypothesized that releases of OPs were responsible for fish and lobster kills in Long Island Sound found in 2001. The persistence in water at 10°C was reported by Muhlmann, et al.[130] Table 8.2 lists half-lives of various OPs.

A suggested treatment involves the use of OP-degrading enzymes. Enzymes are biological catalyst proteins that can be used to detoxify both trace and large quantities

TABLE 8.2
Half-Lives of Organophosphate Pesticides

Compound	Half-Life (Days at 10°C)
Paraoxon	1200
Parathion	3000
Dipterex	2400
Methyl parathion	760

of pollutants. However, because enzymes are proteins, they can be too sensitive to environmental conditions to be of much use. Nevertheless, Orica Corporation of Melbourne, Australia, achieved significant success in field trials intended to remediate OP-contaminated agricultural run-off (aliphatic, nonvinyl, and aromatic OPs).

A cotton farm in Narrabri, New South Wales, was contaminated with irrigation runoff. The 80,000-liter holding pond was treated with a free form of the Orica enzyme. A 90% reduction in methyl parathion was affected within 10 minutes. The final concentration decreased to 0.4 ppb after 1 hour from an initial concentration of 7 ppb.[131]

At a pear and apple orchard in Tatura, Victoria, the rinse water from cleaning equipment was treated. Again, 99% of the methyl parathion was degraded in 1 hour. Figure 8.2 plots methyl parathion and nitrophenol (a product of the degradation) concentrations. The duration shown in the figure is 70 minutes because the half-life of the enzyme is 1 hour. While this method is acceptable for testing small ponds and holding tanks, it would not be an efficient way to continuously treat runoff or large bodies of water.[131]

TECHNIQUES FOR IMMOBILIZATION

Enzymes in solution behave like other solutes in that they have complete motion within solvents. Immobilization is a process by which binding of the enzyme to a substratum eliminates this freedom of movement. If the binding is done at a position on the enzyme that is not used in the catalysis, the activity is maintained.

Several choices must be considered when choosing an immobilization technology. They are remarkably similar to the choices made when choosing a scaffold for cell growth. The immobilization should take place on a material that is strong and tough, has a porous structure, high surface area, permeability, mass transport, and space for biomass buildup. The material should be hydrophilic and inert but have the ability to bind enzymes. It should be hydrolytically stable and be able to resist the environment in which it will have to operate; it cannot be toxic to the enzyme.

Immobilization can be accomplished by several methods. Adsorption is the easiest and least expensive. The bond is weak, however, and loss of enzyme by washout is inevitable. Covalent bonding can be accomplished by activating the

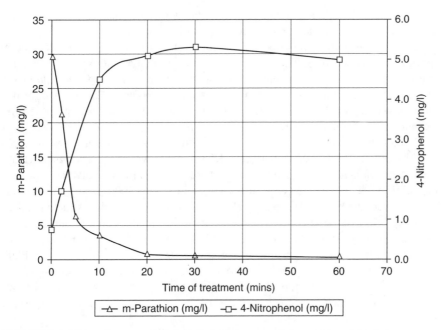

FIGURE 8.2 Effect of organophosphate hydrolase on methyl parathion degradation.

substratum and creating a chemical bond between it and an amine group on the enzyme.

Entrapping an enzyme within a gel is a common technique, but washout and membrane diffusion are important issues. The most effective system would be an enzyme that attaches covalently to a flow-through medium of high surface area to create a system that would represent an optimal balance of both chemical and engineering attributes. This optimal balance is the basis of several applications for polyurethane and its composites.

Polyurethanes offer a convenient method by which immobilization of enzymes can be affected. Prepolymers are polymers with active end groups. While the primary purpose of the isocyanate end groups is serving as chain-extending agents, they also react with the amines that characterize an enzyme backbone. Thus, as many of the studies cited will show, the reaction sequence is (1) preparation of an aqueous solution of the enzyme and (2) emulsification of the solution with the prepolymer. The reaction time is on the order of 0.5 hours compared to the 24 hours required by some methods.

The primary purpose of immobilization is to stabilize the enzyme. Stabilization typically reduces the activity of the enzyme. Part of the reason is stearic, but the reduction of activity could also occur because the site of attachment of the enzyme and substratum is also the portion of the enzyme responsible for its catalytic activity.

LeJeune et al.[132] investigated this stabilizing effect of immobilization at the expense of activity. They immobilized an organophosphorus hydrolase by emulsifying a solution of the enzyme with a hydrophilic polyurethane prepolymer. They intended to study the enzymatically catalyzed hydrolysis of organophosphate nerve gases including soman and sarin. They reported a 100-fold increase in half-life of the enzyme while maintaining over 50% of its native activity.

Havens and Rase[133] reported the immobilization of an enzyme to degrade a specific organophosphate. The organophosphate was an agricultural grade material (parathion). The enzyme was harvested from recombinant *Pseudomonas diminuta* and immobilized by emulsifying a solution with a prepolymer. The product of the reaction was reported to have excellent stability and the method was proposed for cleanup of small spills of parathion.

Beta-galactosidase from *Aspergillus niger* and *Aspergillus oryzae* was immobilized by polyurethane foam made from Hypol 2000 and 3000 prepolymers (Dow Chemical, Midland, MI).[134] Both are TDI-based hydrophilic prepolymers; the difference is a slightly higher cross-link density in Hypol 3000. Using the method taught in the last two examples, the Hypol 3000 showed the highest activity. The immobilization was studied by measurements of the kinetics and by scanning electron microscopy. Immobilization occurred predominantly by covalent bonding between primary amino groups of the prepolymers. Entrapment in the PUR micropores assisted the immobilization of enzymes, while adsorption on the surfaces of macropores was not important. The loading capacity, enzyme activity, and deactivation during immobilization were studied. The immobilized enzyme may be useful for lactose hydrolysis.

In another kinetics study, Huang and Chen immobilized jack bean urease in the form of a thin film on the surface of a reticulated polyurethane foam.[135] The residual apparent activity of the urease after immobilization was about 50%. The good hydrodynamic properties and flexibility of the support were retained in solution after immobilization. Urea hydrolysis was examined in both a batch squeezer and circulated flow reactor. The results suggest potential for practical applications in various reactors.

Storey et al. immobilized glucoamylase with two grades of hydrophilic polyurethane, a foam and a dense hydrogel.[136] The foam was a better support for enzyme immobilization. The enzymes hydrolyzed large (molecular weight over 200,000) substrates as effectively as smaller ones. The long-term storage stability of glucoamylase was enhanced by immobilization in foams (70% activity retained; the free enzyme retained only 50%). Immobilization also improved the enzyme stability to various denaturing agents (sodium chloride, urea, and ethanol). The immobilized enzyme also exhibited better stability at high temperatures compared to the free enzyme.

Storey and Chakrabarti[137] developed a coenzyme system for the conversion of cellulose to fructose. They used a hydrophilic polyurethane foam as an immobilizing substratum and showed that immobilized glucose isomerase had a half-life of about 160 hours during continuous hydrolysis, they observed 35% to 40% activity remaining after 1000 hours. Cellulase, beta-glucosidase, and glucose isomerase were immobilized on hydrophilic foam and cellulose substrates. Glucose production far exceeded fructose production over the initial time points, but the net amounts of fructose produced increased substantially over longer incubation times.

The same researchers studied glucose production using beta-glucosidase immobilized with the same hydrophilic foam (Hypol 2002).[146] The immobilized enzyme
showed 95% retention of activity after 1000 hours of continuous use at 23°C.
Co-immobilization with cellulase yielded a cellulose-hydrolyzing complex with a
2.5-fold greater rate of glucose production for soluble cellulose and a 4-fold greater
increase for insoluble cellulose, compared to immobilized cellulase alone.

It is useful to note that many of the experiments reported here were done on
what might be termed laboratory scale. The sample preparation involved making a
small amount of foam and cutting it into pieces because it is nearly impossible to
pass fluids through a hydrophilic foam. Even under the most carefully controlled
conditions, the best one can hope for is an open structure. The cell structure and
lack of compressive strength combine to "blind off" the matrix if an attempt is made
to pass water through the foam under even modest flow conditions. To avoid a
destructive exothermic reaction (denaturing the enzyme), the foam is made at 4°C.
This process has a tendency to produce higher density, closed-cell foams. Finally,
immobilization involves a partial reaction of the isocyanate functionality with the
enzyme. This takes us further from the ideal of a flow-through hydrophilic foam.
To combat this problem, after immobilization, which evenly distributes the enzyme
throughout the foam, it is common practice to cut the foam into cubes to increase
surface area.

This problem led to the development of the composite we have mentioned earlier.
Grafting the hydrophilic polyurethane onto the reticulated substratum while simultaneously immobilizing the enzyme allows us to maintain temperature control by
passing cool air through the foam. We are not as concerned about cell structure; that
duty is handled by the reticulate.

The literature contains numerous references to comparisons of enzymes immobilized on gel beads and polyurethanes. These immobilizations are surface effects
and unless the data are normalized to the contact area, the comparisons are suspect.
It is clear, however, that immobilization with polyurethane is the easiest of the known
techniques.

Our laboratory has used the composite technology described in Chapter 2 to
produce a scaffold for the immobilization of enzymes. We used reticulated foams
of various sizes and chemistries (polyethers or polyesters) and graft hydrophilic
polyurethanes to their inside structures.

IMMOBILIZATION OF LIPASES ON COFOAM HYDROPHILIC POLYURETHANE

We collaborated with Professor Palligarnai Vasudevan of the Chemical Engineering
Department of the University of New Hampshire on a study of immobilization of
lipases on CoFoam. Immobilization was performed at the Hydrophilix facility in
Portland, ME. Approximately 2 g lipase (from porcine pancreas and *Mucor miehei*)
were stirred into 500 ml deionized water. The enzyme solution was emulsified with
an equal volume of a methylene diisocyanate (MDI)-based hydrophilic polyurethane

TABLE 8.3
Immobilization of Lipase on CoFoam
Hydrophilic Polyurethane

Lipase	Triacetin (U/g biocatalyst)	Tributyrin (U/g biocatalyst)
Porcine Pluronic	1.2	0.55
Mucor miehei	2.4	2.1
Candida antarctica B	255	590

prepolymer (Urepol 1002, Envirochem, Paso Robles, CA). Both liquids were mixed at room temperature. The emulsion was immediately applied to a 30-ppi polyether reticulated polyurethane foam (Rogers Foam, Somerville, MA) and passed through a set of pinch rollers in such a way as to coat the entire inside structure of the foam. The foam was tack-free in 2 minutes and fully cured in 1 hour, after which it was air dried at ambient temperature. Table 8.3 reports the results.

The CoFoam exhibited slightly greater activity toward triacetin and olive oil compared with Novozyme 435. The result with olive oil is especially interesting in relation to applications such as biotransformations of oils and fats. In contrast, the activity with tripropionin and tributyrin was fivefold higher with Novozyme 435 compared with the composites. The extent of leaching was determined by soaking the immobilized enzymes in distilled water at 4°C overnight. The activity of the enzyme was measured again; activity loss was less than 5%.

The immobilization of lipase on CoFoam was considered effective and simple. The low resistance to fluid flow made CoFoam a viable support to be used in large reactors. The enzyme can easily and quickly be loaded in any packed bed reactor.

In another study, catalase derived from bovine liver was immobilized on the same polyurethane composite. The bovine liver catalase enzyme was obtained from City Chemical, West Haven, CT) and used without further purification. The activity of the enzyme was not stated or determined.

The prepolymer used for this study was an MDI-based preparation (Suprasec 1002) from Huntsman PU, Brussels, Belgium. Immobilization of the enzyme was accomplished by coating the inside structure of a reticulated foam with an acetone solution of a hydrophilic polyurethane prepolymer. A 45-pore/in. reticulated foam cut into sheets 0.25 in. thick was used. The coated reticulated foam was immersed in a catalase solution (5 µg/ml) at 4°C and left in the solution for 1 hour to ensure a complete solution. It is known that both the water and some functionality in the catalase react with the isocyanate groups to cause polymerization (the water–isocyanate reaction) and chain termination (the catalase reaction). Controlling the relative concentrations and temperature permits control of the physical properties of the composite and the ability of the foam to function as an enzyme.

The coating weight was a ratio of 3.2 g hydrophilic foam to 1 g reticulated foam. Unlike immobilization on beads or gels, this technique yields a sheet of material of good durability. Solutions were made of 300 ppm hydrogen peroxide from a 30%

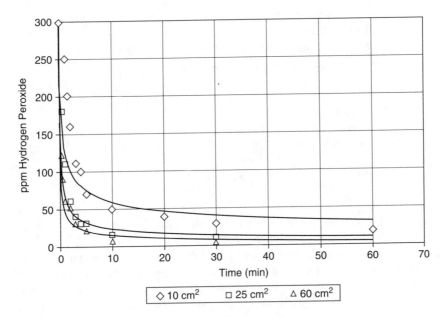

FIGURE 8.3 Degradation of hydrogen peroxide.

stock solution. The concentration of the stock solution was determined by the sodium iodate–thiosulfate titration method. For each determination, a 100.0-ml solution was prepared and placed in a vessel connected to a manometer for measuring the pressure. The vessel was sealed after insertion of a measured piece of catalase-immobilized CoFoam. The reaction of catalase with peroxide produces O_2, and an increase in pressure indicates a degradation of the peroxide. Thus, a change in pressure in the vessel is a measure of the reaction rate. Since it is sufficient to show differences in test samples, the ideal gas law was used to convert the pressure into mass. The barometer was calibrated with a gauge traceable to National Institutes of Standards and Technology (NIST) standards.

The object of this study was to demonstrate the relationship of the surface area of the CoFoam composite and the rate of degradation of the hydrogen peroxide. Accordingly, several pieces of CoFoam of different sizes were placed in the vessel and the pressure monitored with respect to time. Pressure was converted to the amount of hydrogen peroxide degraded. Figure 8.3 reports the data in graphic form.

IMMOBILIZATION OF CELLS

The covalent attachment of enzymes to a polyurethane is not the only method in which enzymes are used. In many cases, it is more convenient to immobilize the cells that produce the enzymes. The extracellular enzymes produced by the cells are then used as we will illustrate below. Immobilization is also important for the production and harvesting of enzymes and other proteins with the objective of increasing the useful lives of organisms. This was shown by Bucke, who immobilized *Erwinia*

TABLE 8.4
Effect of Immobilization Method on Transformation of Sucrose by *Erwinia rhapontici*

Method	Half-Life (h)
Free cells	36
Entrapment in alginate gel	8500
Adsorption on modified cellulose	400
Cross-link with glutaraldehyde	40
Adsorption on bone char	25

rhapontici in various systems.[138] In all cases, the activity of the system was reduced and the half-life was extended (Table 8.4).

Despite the apparent advantages of immobilizations, most enzymes are most commonly produced via a stirred tank reactor (also known as a batch reactor) process. Stirred tank reactors are typically made of stainless steel and range in size from 1-l lab reactors to 500,000-l production scale systems. The reactors are charged with a nutrient solution and inoculated with a microorganism. The microbes produce the extracellular enzymes to digest the nutrients. The cells grow and divide and the amount of enzyme in the medium increases as a function of the amount of cellular material. At the proper time, the medium is removed from the reactor and the cells are separated, yielding a solution of the extracellular enzymes that are purified at least in part by a membrane process. The isolated enzymes are then dried or otherwise prepared for sale. As the name suggests, this process produces a finite amount of enzymes in single batches.

While the batch process is the dominant one in current use, researchers and companies have attempted to create continuous bioreactor systems. Lopez et al. immobilized *Candida rugosa* in polymethacrylamide hydrazide beads and polyurethane foam[139] with the intent to achieve the continuous production of lipase enzymes. Despite flow problems with the polyurethane foam, it showed high lipolytic activity. Biomass buildup was problematic. Feijoo et al.[140] immobilized *Phanerochaete chrysosporium* on polyurethane foam in packed bed bioreactors under near-plug flow conditions. Continuous lignin peroxidase production was accomplished, the rate of which was studied as a function of recycle ratio.

Ariff and Webb[141] studied production of glucoamylase using freely suspended cells of *Aspergillus awamori* in batch and continuous fermentations. Glucoamylase yields based on glucose consumed were 900 and 1080 U/g for batch and continuous fermentations, respectively. The immobilization of viable cells was achieved by adsorption to cubes of reticulated polyurethane foam. In comparison with freely suspended cell fermentations, neither batch nor continuous fermentations of immobilized cells improved glucoamylase production significantly in terms of yield or productivity.

Targonski and Pielecki investigated the production of cellulase using immobilized mycelium of *Trichoderma reesei* mutants on polyurethane foam impregnated with lactose medium.[142] The enzyme yield on lactose was 520 FPU/g of lactose metabolized in comparison with 160 FPU/g using a stirred tank bioreactor.

Rhizopus oryzae was immobilized in polyurethane foam cubes by Sun et al.[143] The effects of the cube size on cell immobilization, cell growth and L(+)-lactic acid production were studied. Immobilization was accomplished by simple adsorption. The use of small cubes for *R. oryzae* immobilization was very effective in increasing the productivity of L(+)-lactic acid by the immobilized cells. The inoculum size was effective for increasing the immobilization ratio (ratio of the number of cubes containing cells to the total number of cubes). We discussed mass transport problems of polyurethane foam and how the application of certain composite technologies mitigates this difficulty earlier in this chapter.

While not specifically related to polyurethanes, a study by Hall and Rao serves as a useful review of the immobilization of algae by various methods.[144] The reasons for using green and red algae and cyanobacteria are reviewed. Possible applications include the production of fuels, fertilizers, biochemicals, and pharmaceuticals; the extraction of heavy metals from aqueous solutions; and urban wastewater treatment.

Domingo et al.[145] evaluated immobilization of trichloroethylene-degrading *Burkholderia cepacia* bacteria using hydrophilic polyurethane foam. The influences of several foam formulation parameters upon cell retention were examined. Surfactant type was a major determinant of retention; a lecithin-based compound retained more cells than Pluronic- or silicone-based surfactants.

Tuerker and Mavituna immobilized *Trichoderma reesei* within the open porous networks of reticulated polyurethane foam matrices. Growth pattern, glucose consumption, and cellulase production were compared with those of freely suspended cells.[146] The method of immobilization was simple and had no detrimental effect on cell activity. Hundreds of similar projects could be cited. Not all rated the use of polyurethane as the preferred technique. If a statistical analysis were conducted on all the immobilization literature, we are sure that no single technique would be dominant. However, the combination of ease of immobilization, cost of materials, flow-through properties, control of flux rate through the immobilizing membrane, high surface-to-volume ratio, and other factors make polyurethane a viable substratum for the continuous production of proteins.

Many of the authors cited above were not specific about the polyurethanes they used. As we know, the chemistry of the polyurethane, its pore size and other factors are important determinants of effectiveness. Some articles did not indicate whether the polyurethanes used were hydrophilic or hydrophobic. One clue we found reliable is that if the enzyme of a cell is mixed with a prepolymer, more often than not the polyurethane is hydrophilic. A reticulated foam used for immobilization is typically a hydrophobic polyurethane. It is hoped that this book will influence the research community to be specific about the chemistries of their scaffolds.

IMMOBILIZATION STUDIES: SUMMARY

This introduction of the use of polyurethane for the immobilization of enzymes and cells indicates that both the chemistry and the physical structure of polyurethane make it appropriate for consideration as a commercial scaffolding material. While we have not discussed costs, the commodity nature of the reticulated substratum and the availability of hydrophilic polyurethane prepolymers could make polyurethanes the preferred materials for large-scale operations.

USE OF HYDROPHILIC POLYURETHANE
FOR CONTROLLED RELEASE

We discussed the four primary attributes of a polyurethane (architecture, reservoir, biocompatibility, and attachment of ligands) in earlier chapters. Biocompatibility and ligand attachment were discussed in detail, as was architecture in relation to the other attributes. This section covers reservoir capacity and cosmetic delivery applications. Fragrance, soap, pharmaceutical, and other delivery applications are other practical uses for polyurethanes.

Delivery applications are based on the fact that polyurethanes swell when exposed to solvents. Once a material is imbedded in a polyurethane matrix, the matrix can deliver the solvent or active ingredient in a number of ways including solubility and vapor pressure. Other polymer systems have this ability, but the versatility of the polyurethane molecule (e.g., hydrophilicity or hydropobicity) gives it special properties. The ability to produce polyurethanes as foams or elastomers grants product designers additional degrees of freedom.

SKIN CARE DELIVERY APPLICATION

Our laboratory partnered with Joséphine Bissing Dermaceuticals of Denver, CO, to develop delivery systems for certain skin care formulations that would eliminate the needs for surfactants, stabilizers, and emulsifiers. In effect, the hydrophilic polyurethane matrix in which the skin care formulation resides would replace all the stabilizing chemicals typically used in such formulations.

The skin is the largest organ of the body. It functions as more than a covering. Skin is a multifunctional device; it regulates heat, protects, and neutralizes harmful environmental factors. Skin also serves as a sensory organ that allows the rest of the body to respond to touch. It is usually the first thing people notice.

Clarity, suppleness, and elasticity characterize skin during youth, due in part to continuous regeneration as new skin cells replace old ones. As people age, however, this regeneration ability slows. In addition, chemical and physical insults and other factors can change the texture, smoothness, and even the appearance of skin.

Science has made remarkable gains in the area of skin care in the past 20 years. The development of vitamin A derivatives brought clinically proven reductions in wrinkles. Fruit acids, Ceramide compounds and a library of plant derivatives diminished the impacts of aging.

An anecdote is useful at this point. During development meetings at a large skin care marketer, we asked how the company would propose to market a product if we had an ingredient that showed remarkable properties in clinical trials. All the 15 or so senior product design chemists in the room wanted to build the ingredient into a cream or a lotion. We mention this not to suggest that creams and lotions are not viable methods of delivery but to make a point that they should not be the only methods. Manufacturers are beginning to see the point of our question, however. In the past 2 years, several serum-based formulations that do not contain the formularies of typical creams have become available.

A stable emulsion requires balancing of hydrophobic and hydrophilic components into a uniform suspension. The suspension may be a multicomponent system of emulsifiers, surfactants, stabilizers, and thickeners that are included only to make an aesthetically pleasing product — not for their clinical significance.

The object of the Joséphine Bissing group is to eliminate as many of these nonactive ingredients as possible and incorporate the active ingredient directly into the matrix of a hydrophilic polyurethane.[147] It is still necessary to add a diluent to the product to reduce its activity, but the hydrophilic polyurethane provides the freedom to use ingredients that play moisturizing or other functional roles. Typically, the diluent is a plant-derived glycerine or in some cases mineral oil.

CLINICAL STUDIES

Prospective, multisite, uncontrolled clinical studies were conducted to study the effectiveness of the formulations and the acceptability of the polyurethane delivery system. The formulation was imbibed into 2.5-in. hydrophilic polyurethane foam discs (LMI, St. Charles, MI). Each disc contained 2.4 g of the formulation. The method of application was developed in tests that limited the active ingredient to the same amount or less than the monograph specifications for the formulation ingredients. Fourteen volunteers were given boxes containing 42 individually wrapped foam pads impregnated with the formulation and were asked to complete the questionnaire weekly.

Inclusion and Exclusion Criteria

Trial participants had to be 30 years of age or older and their skin had to show wrinkles, thinning, dryness, or other superficial signs of aging. Bases for exclusion were unwillingness to follow study protocol guidelines, unwillingness to sign an informed consent, and concurrent use of products seeking to address the same purposes as the product tested.

Instructions to Participants

Trial participants were given the following instructions:

1. Each evening for the 42 days of the trial, thoroughly clean your face. Follow cleansing with a pH-balanced toner.
2. Remove a pad from the foil packaging.
3. Moisten your face with water. Lightly but firmly pat the pad over the entire face, particularly around the mouth and eye area.
4. Each pad contains enough active ingredient to apply the treatment to other areas at your discretion.

At the start of the study, each participant was asked to evaluate the condition of facial skin using a prestudy evaluation sheet. At the end of each 7-day period after the beginning of the test, each participant was to use a magnifying mirror to examine his or her face and answer questions on an evaluation sheet. At the end of the study, the questionnaires were compiled and analyzed to determine trends. As important as the questionnaires were the positive and negative comments of the participants concerning their impressions of improvements.

Results

The parameters of interest were:

- Moisturization
- Softness
- Suppleness
- Texture
- Firmness
- Hydration
- Lines and wrinkles
- Effectiveness of foam pad delivery system

The self-evaluation was based on a scale of 1 to 5. A proposition was stated. If a respondent strongly agreed, he or she noted a 5; if he or she strongly disagreed, the response was 1. The evaluation of the formulation was very positive; the average score was 4.8. Most important were the poststudy comments in which participants rated the sponge delivery system as very good to excellent.

AGRICULTURAL APPLICATIONS

There is no more appropriate application of the control of structure and chemistry than the use of polyurethane in agriculture. The leader in this technology is International Horticultural Technologies in Hollister, CA. For the past two decades, this company has promoted a composite of hydrophilic polyurethane and natural fibers for the propagation of high-value plants. Research at various universities has shown the applicability of this technology to ornamentals and other woody plants. Our laboratory suggested use of the composite in silviculture based on the stability of the growing media both as a way to shorten greenhouse time and enhance the ability to transplant the trees with automatic equipment. All these applications would lead to a higher survival rate during transplantation because the root system would be protected by the polyurethane plug.

It is not the purpose of this book to define agricultural applications in detail because the subject is broad and any discussion would have to include plant-specific issues. It is, however, appropriate to describe design procedures and discuss the uniqueness of polyurethane in agricultural products.

The most important design feature of an agricultural growing medium is a balance of void volume and moisture content. The moisture tension test is the most

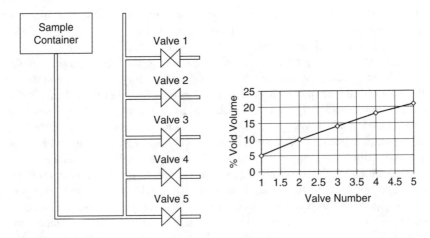

FIGURE 8.4 Apparatus for determining moisture tension.

effective way to determine this balance. Figure 8.4 is a schematic of the technique. A sample of the growing media is fully saturated with water, that is, the matrix is fully saturated and the void volume is filled with water. For most growing media, we have found that the density of the sample is only slightly less than the density of water. The sample is then allowed to drain. As shown in the figure, the drain tubes are placed at levels below the bottom of the sample container and the amount of water leaving the sample is determined. Based on the density of water and this apparatus, one can calculate the void created by the drained water. Plotting the void volume vs. the position of the drain tube relative to the bottom of the sample container provides an indication of how the material will fare as a growing medium.

We have defined the properties faced by developers of new growing media. Based on these definitions, it is clear that reticulated foam would not be useful for such applications. Even at the finest pore size, virtually no water is retained during this test. If, however, we were to graft hydrophilic polyurethane onto a reticulated foam, the grafted polyurethane increases the amount of water retained in the moisture tension test. If an organic fiber such as peat moss is included in the hydrophilic polyurethane, the effect is even more pronounced. Applications in both hydroponic and soil planting are anticipated from this technology.

ARTIFICIAL MUSCLE DEVELOPMENT

Since the early 1950s, researchers have been intrigued by the idea of producing a material that would mimic the action of a muscle; that is, a system that would contract or relax upon demand. Using the human muscle as a model, the most commonly expressed goal was to establish a system that would respond to an electrical signal.

A synthetic muscle system would have a broad range of uses. A recent article[148] hypothesized that the future of heart restoration would be the replacement of a damaged heart chamber with a synthetic muscle. Implantable circulatory-assist devices would take over some of the burdens on damaged hearts, and injured or diseased muscles in other parts of the body could be replaced.

Applications in industry would be equally broad. Chief among them would be lightweight robotic equipment that did not require large and heavy power supplies.

Synthetic muscle systems were first mentioned by Flory[149] in 1953 and defined in the context of this chapter by Kuhn et al.[150] Flory described the effect of temperature on a polyacrylic acid–polyvinyl alcohol blend. Kuhn studied pH, temperature, and ionic strength stimuli. Tanaka et al.[151] made a significant contribution to the technology in an experiment that combined the contraction–relaxation phenomenon with stimuli developed indirectly by an electric current. Their work resulted in a patent.[152]

Concurrent with the use of hydrogels in this application was a radically different direct approach to the same goal: a material that would contract and relax upon demand. Specially prepared electrically conductive polymers undergo structural changes that result in contraction. A recent article described what the authors proposed as the best of this class of device.[153] Carbon nanotube sheets were treated as electrolyte-filled electrodes. When a current passed through an apparatus constructed with this material, the tubes bent according to the direction of the current flow. The data presented in the paper show fast and strong responses. The degree to which the material contracts, however, is not sufficient for what are normally considered primary uses, i.e., robotics and artificial muscles. The researchers reported cantilevered displacements on the order of fractions of millimeters for a 20-mm film that are insufficient for most applications, in our opinion. Nevertheless, the direct conversion of electrical energy to mechanical energy has some attractive advantages unmatched by hydrogel systems that we would propose.

When a water-soluble polymer is dissolved in water, a complex network is formed that includes the polymer backbone, free water, and water in various degrees of bonding to the polymer. Depending on the concentration of polymer, its molecular weight, and several other factors, the network of polymer and bound water can assume the volume of the solution. This, of course, leads to the high viscosity that these solutions develop. The volume occupied by the polymer and the associated water in the system are said to be the hydrodynamic volume. As this volume increases because of increases in molecular weight or in the water shell surrounding the molecule, the viscosity of the solution increases.

A number of polymers exhibit this hydration property. Natural products such as cellulose and starch are or can be made water soluble. Synthetics such as polyvinyl alcohol and polyacrylic acid are also soluble in water. This discussion will be limited to synthetic materials such as polyacrylic acid and its salts, polyvinyl alcohol, polyacrylamide, and polyurethane

All but the polyurethane are characterized by methylene backbones with ligands that are sufficiently polar to make them water soluble. Thus, upon dissolution in water, the polarity of the water molecule associates with the polarity of the acrylic or acrylamide groups to form a shell. We discussed hydrophilic polyurethanes that are typically cross-linked and are not (but could be) considered effective thickeners. Nevertheless they too have hydration shells developed due to the influence of the polyethylene glycol backbone. The extent of that shell is determined by the hydrophilicity of the ligand; the acrylic > acrylamide > alcohol > polyurethane. The volume of the shell around the ligand is largest around the acrylic. It follows, therefore, that the acrylic would be the most effective thickener because the lowest concentration of the acrylic would be required to achieve a given viscosity. Other factors affect hydrodynamic volume, however, and they serve as the foci of researchers seeking to develop artificial muscles from hydrogel materials.

It is necessary to discuss another chemical feature related to water-soluble polymers: cross-cross-linking — the component that separates viscous systems from gel systems. Viscous systems flow, and it follows, therefore, that they do not possess the tensile properties of muscles. High-viscosity systems have structural integrity, gels provide the necessary combination of tensile strength and elongation or stretch.

One important factor that differentiates the acrylic polymers from the other three: they are ionic and the others are nonionic. This difference is important because nonionic polymers do not respond to changes in ionic strength and ionic polymers do.

A polymer hydrogel used in the development of an artificial muscle must have strength. The saying that "a chain is only as strong as its weakest link" certainly applies. Briefly, a polymer is used for its hydrophilicity and the cross-linker is used to adjust the physical properties of the system. Increasing the cross-linking increases tensile strength and decreases the elongation or stretchiness. A balance of both properties must be established to address the specific needs of an application. It is also a general rule that hydrophilicity decreases as cross-linking increases because the cross-linking blocks a portion of the backbone and thereby inhibits the free development of a water shell around the polymer. Therefore, while cross-linking is an essential component of a polymer system, the way it affects hydrodynamic volume must be considered.

The fundamental property at work is the interaction of water and polymer. In the Kuhn experiments, the acrylic acid functioned as the contractile unit.[159] A change in pH or ionic strength will either hydrate or dehydrate a gel. This affects the size of the molecule, which in turn causes the molecule to contract or expand. This phenomenon is most pronounced in ionic polymers.

In summary, we described a system in which a backbone containing a highly polar group contains a degree of cross-linking. Put another way, the system is a three-dimensional network of water-soluble polymer and cross-linking that serves as the basis for all hydrogels, natural or synthetic.

A hydrogel placed in an excess of water will absorb the liquid until it reaches a maximum. This ability is typically reported as the percent water in a fully swollen gel. The hydrophilicity of the polymer and the degree of cross-linking determine the degree to which the gel will absorb. Some hydrogels contain as much as 99% water. An acrylic acid gel will have a higher equilibrium moisture than a polyvinyl alcohol gel. This characteristic is not unlike the hydrodynamic volume factor described above.

As the equilibrium moisture increases, however, the strength and elasticity of the hydrogel will decrease. Thus, a balance between hydrophilicity and physical strength must be established, and the balance is critical to the development of an artificial muscle.

The strength of a gel is based on a combination of molecular weight, degree of cross-linking, and the hydrodynamic volume of the gel equilibrium moisture. By including a cross-linking component in a water-soluble polymer and adding it to water, the system will gel and in effect develop infinite viscosity. The degree of cross-linking affects the strength of the gel. Most importantly, the responses of ionic polymers to their environment (pH, ionic strength, temperature, etc.) are the driving force for contraction.

GEL PREPARATIONS

This section will report on our studies of the contractile properties of several polymers and our work with a composite of gels with specific functions:

- A nonionic polyurethane gel that serves as a reservoir for contractile ions
- A cationic gel that contracts under the influence of anions
- An anionic gel that contracts under the influence of cations

POLYURETHANE HYDROGEL

A hydrophilic polyurethane prepolymer was made according to the procedure taught by Braatz.[163] The prepolymer was mixed with water at a ratio of 10 parts water to one part prepolymer. The emulsion was poured immediately onto a silicone release liner and allowed to cure for 30 min. It was free of voids and had a density roughly the same as water. For the contraction experiments, the gel was immersed in excess distilled and sterile water for 24 h, then cut with a steel-ruled die into circles 40 mm in diameter.

CROSS-LINKED POLYACRYLAMIDE GELS

Premixed blends of acrylamide and bisacrylamide prepared with varius ratios of monomers were purchased from Eastman Kodak Chemical Company (Rochester, NY). The 37.5:1 and the 19:1 preparations were used for the study. Gels made from these mixtures will be referred to as 2.6% and 5% cross-linked polyacrylamides, respectively. Five grams of each monomer blend were added to 95-g portions of distilled water. Solution was achieved by mixing for 1 h. To each sample was added 1 ml each of a 1% solution of N',N',N',N'-tetramethylethylenediamine (Eastman Chemical Co., New Haven, CT) and a 10% solution of ammonium persulfate (Mallinckrodt Laboratory Chemicals, Phillipsburg, NJ). The solutions were poured into an open polyethylene mold and allowed to cure for 12 h at room temperature. The gels were carefully removed and placed in an excess of distilled sterile water for 48 h. The water was replaced several times during the equilibration period. It was felt that this was sufficient to remove unreacted monomers and impurities. The gels were then cut with a steel-ruled die into circles 40 mm in diameter.

CROSS-LINKED POLYACRYLIC ACID GELS

Polyacrylamide gels of the two cross-linking levels mentioned above were prepared and cut by the same procedure. The circles were placed in a 1N sodium hydroxide solution. Conversion of the amide groups to the carbonyl was evidenced by the evolution of ammonia. The diameters of the gels were monitored over a period of 3 days until they stabilized (indicating near complete conversion to the neutralized acrylic acid form). The circles were then transferred to a distilled water bath where they remained for 72 h with frequent changes of water. The circles were then die-cut to 40 mm in preparation for the contraction experiments.

CONTRACTION EXPERIMENTS

Each of the gels was soaked in deionized water for 24 h, and then in salt solutions. The degree to which they responded to the salt represented a measure of efficacy as an artificial muscle. Figure 8.5, Figure 8.6, and Figure 8.7 report the responses of three of the polymer systems to various ions.

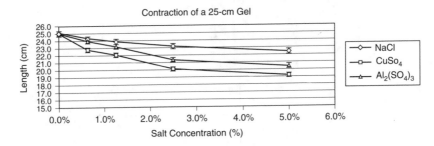

FIGURE 8.5 Effects of ions on polyacrylic acid gels.

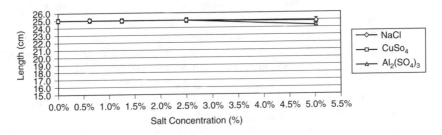

FIGURE 8.6 Effects of ions on polyacrylamide gels.

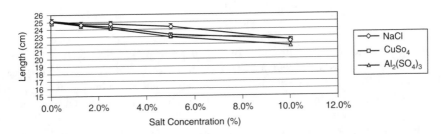

FIGURE 8.7 Effects of ions on polyurethane gels.

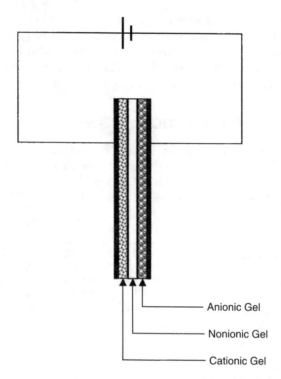

Anionic Gel

Nonionic Gel

Cationic Gel

FIGURE 8.8 Composite of a nonionic polyurethane core of an artificial muscle.

From these and other experiments a construction was developed that included a core of nonionic hydrophilic polyurethane. The core served as a reservoir for ionized salt. A layer of anionic gel was placed on one side of the core; a layer of cationic gel was placed on the opposite site. A current passed through the gel caused the cation to migrate into the anionic gel and led the gel system to contract. When the current flow was reversed, the ions returned to the reservoir.

We found that by using a square wave generator, the flow of ions in the desired direction was accomplished by increasing or decreasing the duration of the wave in the plus or minus position. Embedding a strain gauge in the gel and using a feedback algorithm allowed us to achieve precise control of the contractions. Figure 8.8 shows the circuitry.

CONCLUSION

Our intent was to encourage invention by showing how polyurethane can participate in the design process as a specialty chemical in its own right. As a foam, it has a high surface area and a low void volume of significant strength. By using polyurethane's component ingredients and reservoir capacity, protein resistance and cell binding can be developed.

It is our hope that this book will serve as the beginning of a successful journey involving new applications of this unique chemistry.

References

1. Squillance, P.J., Morgan, M.J., Lapham, C.V., et al, Volatile Organic Compounds in Untreated Groundwater of the United States, 1985-1995, Environmental Science & Technology, 33(23): 4176-4187, 1999.
2. Jauregui, H., Hayner, N., Solomon, B., and Galletti, P. Hybrid Artificial Liver, in *Biocompatible Polymers, Metals, and Composites.* Szycher, M., Ed., Technomics Publishing, Lancaster, PA, 1983, chap. 39.
3. United Network for Organ Sharing, Annual Report. www.unos.org, 1999.
4. Adapted from Greisheimer, E.M., *Physiology and Anatomy*, JB Lippincott Company, 1950, chap. 16.
5. Matsushita, T., Ijima, H., Koide, N., Funatsu, K., High albumin production by multicellular spheroids of adult rat hepatocytes formed in the pores of polyurethane foam. *Appl. Microbiol. Biotechnol.* 1991. vol. 36, no. 3, pp. 324-326.
6. Jonathan G. Huddleston, Heather D. Willauer, Scott T. Griffin, and Robin D. Rogers, Aqueous Polymeric Solutions as Environmentally Benign Liquid/Liquid Extraction Media, Ind. Eng. Chem. Res. **1999**, 38, 2523-2539.
7. Saunders, J., Frisch, K., *Polyurethanes: Chemistry and Technology, Parts I, page 3,* Interscience Publishers NY, 1967.
8. Bayer, O., Polyurethanes, *Modern Plastics* 1947, 24, 149-152.
9. Product Literature, High Performance Isocyanates for Polyurethanes, The Dow Chemical Company, Midland, Michigan, USA, Form No. 109-799-86, 1986.
10. Heiss, H., Saunders, J., Morris, M., Davis, B., Hardy, E., Polyurethane Adhesives, *Ind. Eng. Chem.* 46. 1498 (1954).
11. Oertel, G., *Polyurethane Handbook*, 2nd ed., Hanser Publishers, Munich 1994.
12. Saunders, J., Frisch, K., *Polyurethanes: Chemistry and Technology, Parts I and II,* Interscience Publishers NY, 1967.
13. Foamex Corporation Product Literature FS-998-F-5, 1999, Eddystone, PA 19022.
14. Thomson, T., US Patent 6,617,014, Composite Polyurethane, 2003.
15. *The ICI Polyurethanes Book*, Ed. Wood, G., John Wiley & Sons, New York, p24, 1987.
16. Adapted from *The ICI Polyurethanes Book*, Ed. Wood, G., John Wiley & Sons, New York, p35, 1987.
17. Foamex Corporation Product Literature FS-998-F-5, Eddystone, PA 19022, 1998.
18. Thomson, T., *Hydrophilic Polyurethanes*, CRC Press, 2000, chap. 6.
19. David, D.J., and Staley, H.B., *Analytical Chemistry of Polyurethanes,* Robert E. Krieger Publishing, Huntington, NY, 1979, pp301.
20. Ibid, pp 303.
21. Ibid, pp 359.
22. Adapted from Saunder, J.H. and Frisch, K.C., *Polyurethanes Chemistry and Technology,* Part I, Interscience,1962 pp 288.
23. Szycher, M., Structure Property Relations, Presented at the 12th Annual Seminar on Advances in Medical-grade Polyurethanes, Technomics Publishing, Lynnfield, Ma, 1998.
24. Saunders, J.H., Frisch, K.C., *Polyurethanes Chemistry and Technology,* Part I, Interscience,1962 pp 253.

25. Adapted from Saunders, Ibid, p 319.
26. Hypol Product Literature, Dow Chemical USA, 1986.
27. Dr. Micheal Szycher, Personal Communication, 1998.
28. Braatz, J.A., Heifetz, A.H., and Kehr, C.L. *J Biomater Sci Ed.* 1992; 3(6):451-462.
29. Merrill, E.W., Salzman, E.W. *J. Am. Soc. Artif. Int. Org.* 1983;6:60.
30. Nagaoka, S., Mori, Y., et al. *J. Am. Soc. Artif. Int. Org.* 1987;10: 76.
31. Jeon, S.I., Lee, J.H., Andrade, and DeGennes, P.G. *J. Colloid Int. Sci.* 1991;142:149.
32. Jeon, S.J. and Andrade, J.D. *J. Colloid Int. Sci.* 1991;142:159 (1991).
33. Amiji, M., and Park, K., Prevention of Protein Adsorption and platelet adhesion of surfaces by PEO/PPO/PEO triblock copolymers, *Biomater.* 13 (10) 1992: 682-92.
34. Faudree, T.L., US Patent No. 4,230,566.
35. Storey, K.B., Duncan, J.A., Chakrabarti, A.C. Immobilization of amyloglucosidase using two forms of polyurethane polymer, *Appl-Biochem-Biotechnol.* 1990 Mar; 23 (3):221-36.
36. A Review of Contaminant Occurrence in Public Water Systems, EPA Report No. 816-R-99-006, Nov. 1999.
37. Lesley-Grady, C.P., Daigger, G.T., and Lim, H.C., *Biological Wastewater Treatment,* Marcel Dekker, 1999.
38. Odabasi, M., Vardar, N., Sofuiglu, A., et al. Polycyclic aromatic hydrocarbons in Chicago air, *The Science of the Total Environment* 227 (1999) 57-67.
39. Rosell, A., Gomez-Belinchon, J.I., and Grimalt, J.O., Gas Chromatographic-Mass Spectrometric Analysis of Urban-Related Aquatic and Airborne Volatile Organic Compounds: Study of the Extracts Obtained by Water Closed-Loop Stripping and Air Adsorption with Charcoal and Polyurethane Foam, *Journal of Chromatography. Biomedical Applications,* Vol. 562, p 493-506, 1991.
40. United Nations Children Fund, *State of the World's Children,* 1993.
41. *Perry's Chemical Engineers Handbook,* Perry, R.H., Green, D.W., and Maloney, J.O., Eds. 7th ed. McGraw Hill, NY, 1997.
42. Gesser, H.D., Spalding, A.B., Chow, A., and Turner, C.W., The monitoring of organic matter with polyurethane foam, J. Am. Water Works Assn., 65, 220, 1973.
43. Saxena, S.R., Kozuchowski, J., and Basu, D.K., Monitoring of polyneuclear aromatic hydocarbons in water, *Environ. Sci. Technol.* 11, 682, 1977.
44. El Shahawi, M.S., Kiwan, A.M., Al-Daheri, S.M., Saleh, M.H., The retention behavior and separation of some soluble organophosphate insecticides on polyester-based polyurethane foams, *Talanta* (42) 1995, 1471-1478.
45. Woese, M.C., Kandler, O., and Wheelis, M.L., Toward a natural system of organisms: Proposal for the domains of Archaea, Bacteria and Eukarya. Proceedings of the National Academy of Science USA 87,4576-4570, 1990.
46. Devinny, J.S., Deshusses, M.A., Webster, T.S., *Biofiltration for Air Pollution Control,* Lewis Publishers, 1999, chap. 9 pp 248.
47. Gabriel, D., Cox, H.J., Brown, J., Biotrickling filters for POTWs air treatment: Full-scale experience with a converted scrubber, *Odors and Toxic Air Emissions,* 2002.
48. Cole, D.C., Evaluation and testing of an experimental biofilter medium, Master of Science Thesis in Marine Bio-resources, August, 1999.
49. Givens, S.W., and Sacks, W.A., Evaluation of carbon impregnated polyurethane foam media for biological removal of carbon and nitrogen from chemical industry waste-water, Proceedings of the 42nd Industrial Wastewater Conference, Purdue University, 1987.
50. Muir, J.F., Recirculated Water Systems in Aquaculture. *Recent Advances in Aqua-culture,* London, 1982.

51. Unpublished report by Deschusse, M.A. 2002. Testing of a New Foam for Biotrickling Filters for Air Pollution Control. Draft & Final Report to Hydrophilix Inc. April 2002. 41 pages.

52. Fava, F., Baldoni, F., Marchetti, L., and Quattroni, G., A bioreactor system for the mineralization of low-chlorinated biphenyls, Process-Biochem. 1996 vol. 31, no. 7, pp. 659-667.

53. Zaiat, M., Vieira, L.G.T., and Foresti, E., Spatial and temporal variations of monitoring performance parameters in horizontal-flow anaerobic immobilized sludge (HAIS) reactor treating synthetic substrate, Water-Res. 1997 vol. 31, no. 7, pp. 1760-1766.

54. Zaiat, M., Vieira, L.G.T., and Foresti, E., Liquid-phase mass transfer in fixed-bed of polyurethane foam matrices containing immobilized anaerobic sludge, Biotechnol. Tech. 1996 vol. 10, no. 2, pp. 121-126.

55. Sun, Y., Li, Y.-L., Yang, H., Bai, S., and Hu, Z.D., Characteristics of immobilized Rhizopus oryzae in polyurethane foam cube, Biotechnol. Tech. 1996 vol. 10, no. 11, pp. 809-814.

56. Borja, R. and Banks, C.J., Kinetic study of anaerobic digestion of fruit-processing wastewater in immobilized-cell bioreactors. Biotechnol. Appl. Biochem. 1994 Aug;20 (Pt 1): 79-92.

57. Vieira, L.G.T., Zaiat, M., and Foresti, E., Intrinsic kinetic parameters of substrate utilization by anaerobic sludge along the horizontal-flow anaerobic immobilized sludge (HAIS) reactor, Environ. Technol. 1997 vol. 18, no. 9, pp. 953-957.

58. Vieira, L.G.T., Zaiat, M., Foresti, E., and Hokka, C.O., Estimation of intrinsic kinetic parameters in immobilized cell systems for anaerobic wastewater treatment, Biotechnol.-Tech. 1996 vol. 10, no. 9, pp. 635-638.

59. Sanroman, A., Pintado, J., and Lema, J.M., A comparison of two techniques (adsorption and entrapment) for the immobilization of Aspergillus niger in polyurethane foam, Biotechnol. Tech. 1994 vol. 8, no. 6, pp. 389-394.

60. Nemati, M. and Webb, C., Effect of ferrous iron concentration on the catalytic activity of immobilized cells of Thiobacillus ferrooxidans, Appl. Microbiol. Biotechnol. 1996 vol. 46, no. 3, pp. 250-255.

61. Fynn, G.H. and Whitmore, T.N., Colonization of polyurethane reticulated foam biomass support particle by methanogen species, Biotechnol. Lett. 1982. vol. 4, no. 9, pp. 577-582.

62. Rao, K.K. and Hall, D.O., Photosynthetic production of fuels and chemicals in immobilized systems., Trends Biotechnol. 1984. vol. 2, no. 5, pp. 124-129.

63. Bailliez, C., Largeau, C., Casadevall, E., Lian, Wan-Yang, and Berkaloff, C., Photosynthesis, growth and hydrocarbon production of Botryococcus braunii immobilized by entrapment and adsorption in polyurethane foams., Appl. Microbiol. Biotechnol. 1988. vol. 29, no. 2-3, pp. 141-147.

64. Brash, J.L. and Uniyal, S., Dependence of albumin-fibrinogen simple and competitive adsorption on surface properties of biomaterials, J. Polymer Sci., C66, 377-389 1979.

65. Waugh, D.F., Anthony, L.J., and NG, H., The interactions of thrombin with borosiicate glass surfaces, J. Biomed. Mater. 9, 511-536, 1978.

66. Chuang, H.Y.K., King W.F., and Mason, R.G., Interaction of plasma proteins with artificial surfaces: protein adsorption isotherms, J Lab. Clin. Med., 92, 483-496, 1978.

67. Phaneuf, M.D., A novel ionic polyurethane with custom-tailored surface Properties, 14th Annual Seminar on Advances in Medical-Grade Polyurethanes, Technomics Publishing, Somerville, Ma, 2000.

68. Szycher, M., Biostability of polyurethanes, 14th Annual Seminar on Advances in Medical-Grade Polyurethanes, Technomics Publishing, Somervillem Ma, 2000.

69. Andrade, J.D., *Surface and Interfacial Aspects of Biomedical Polymers,* Vol. 2 page 1, Pellum Press, NY 1985.

70. Braatz, J.A., Heifetz, A.H., and Kehr, C.L., A new hydrophilic polyurethane for biomaterial coatings with low protein adsorption, *J. Biomater. Polymer Edn.,* Vol. 3, No. 6, pp51-462 (1992).

71. Braatz, J. and Kehr, C. US patent No. 4,886,866, Contact lenses based on biocompatible polyurethane and polyurea-urethane hydrated polymers, 1989.

72. P A Gunatillake, P.A., and Adhikari, R., *Biodegradable Synthetic Polymers for tissue engineering,* European Cells and Materials Vol. 5. 2003 (pages 1-16).

73. Marans, Nelson, S., Pollack, and Alan, R., Biodegradable hydrophilic foams and method, US Patent 4,132,839, 1976, Assignee: W. R. Grace & Co. (New York, NY) 1979.

74. Woodhouse, K.A. and Skarja, G.A., Biodegradable polyurethane, US Patent 6,221,997, 2001.

75. Storey, R.F., Wiggins, J.S., Mauritz, K.A., and Puckett, A.D., Bioabsorbable composites. II: Nontoxic, L-lysinebased (polyester-urethane) matrix composites. Polymer Composites **14**: 17, 1993.

76. Bruin, P., Smedinga, J., Pennings, A.J., and Jonkman, M.F., Biodegradable lysine diisocyanate-based poly(glycolide-co-ε-caprolactone)-urethane network in artificial skin. *Biomaterials* **11**: 191-295, 1990.

77. Zang, J.Y., Beckman, E.J., Piesco, N.P., and Agrawal, S., A new peptide-based urethane polymer: synthesis, biodegradation, and potential to support cell growth *in-vitro. Biomaterials* **21**: 1247-1258, 2000.

78. Hirt, T.D., Neuenschwander, P., and Suter, U.W., Synthesis of degradable, biocompatible, and tough blockcopolyesterurethanes. *Macromol. Chem. Phys.* **197**: 4253-4268, 1996.

79. De Groot, J.H., De Vrijer, R., Pennings, A.J., Klompmaker, J., Veth, R.P.H., and Jansen, H.W.B., Use of porous polyurethanes for meniscal reconstruction and meniscal prosthses. *Biomaterials* **17**: 163-173.

80. Spaans, C.J., Belgraver, V.W., Rienstra, O., De Groot, J.H., Veth, R.P.H., and Pennings, A.J., Solvent-free fabrication of micro-porus polyurethane amide and polyurethane-urea scaffolds for repair and replacement of the knee-joint meniscus. *Biomaterials* **21**: 2453-2460, 2000.

81. Frisch, S.M., and Ruoslahti, E., Integrins and anoikis, *Curr. Opin. Cell Biol.* 9, 1997, 701-706.

82. Hohner, H.P., and Denker, H.W., The role of cell shape for differentiation of choriocarcinoma cells on extracellular matrix, *Exp. Cell Res.* 215: 1994, 40-50.

83. Mescher, M.F., Surface contact requirements for activation of cytotoxic T lymphocytes. *J Immunol.* 149, 1992, 2402-2405.

84. Trudell, L., Thomson, T., Naik, S., Jauregui, H., Laboratory Experience with a New Biomaterial for Covering Wounds and Burns (as well as for other Biomedical Applications, *Symposium on Advanced Wound Care and Medical Research Forum on Wound Repair,* New Orleans, April, 1997.

85. Matsushita, T., Ijima, H., Koide, N., and Funatsu, K., High albumin production by multicellular spheroids of adult rat hepatocytes formed in the pores of polyurethane foam. *Appl. Microbiol. Biotechnol.* **36** 324, 1991.

86. Tomonobu, G., Shimada, M., Shirabe, K., Nakazawa, K., Ijima, H., Matsushita, T., Funatsu, K., Sugimachi, K., Evaluation of a Hybrid Artificial Liver Using a Polyurethane Foam Packed-Bed Culture System in Dogs, *Journal of Surgical Research* **82**, 131-136 (1999).

87. Langer, R., Vacanti, J.P., Tissue Engineering, *Science* **260**, 920, 1993.

88. Cima, L., Vacanti, J., Vacanti, C., Tissue Engineering by Cell transplantation using Degradable Polymer Substrates *J. Biomech. Eng.* **113**, 143, 1991.

89. Allen, J.W., and Bhati, S. N., Engineering liver therapies for the future, *Tissue Eng.* Vol 8, No. 5, 2002, 725-737.

90. Alter, MJ., Epidemiology of Hepatitis C American Liver Foundation, http://www.liverfoundation.org /html/livheal./dir/livheal.htm. 2001.

91. Starzl, T.E., Marchioro, T.L., Kaulla, K.N., Hermann, G., Brittain, R.S., and Waddell, W.R., Homotransplantation of the liver in humans. *Surg. Gynecol. Obstet.* **117**, 659, 1963.

92. Ghobrial, R.M., Yersiz, H., Farmer, D.G., Amersi, F., Goss, J., Chen, P., Dawson, S., Lerner, S., Nissen, N., Imagawa, D., Colquhoun, S., Arnout, W., McDiarmid, S.V., and Bsutill, R.W. Predictors of survival after in vivo split liver transplantation: Analysis of 110 consecutive patients. *Ann. Surg.* **232**, 312 2000.

93. Raia, S, Nery, JR, Mies, S, Liver transplantation from live donors, *Lancet* **2**, 497, 1989.

94. Keefe, EB, Liver transplantation: Current status and novel approaches to liver replacement. *Gastroenterol.* **120**, 749, 2001.

93. Demetriou, AA Clinical experience with a bioartififial liver in the treatment of severe liver failure: A phase I clinical trial- Discussion, *Ann. Surg.* **225**, 493, 1997.
     Runge, D., Runge, D., Jager, D., Lubecki, K., Stolz, D., Karathanasis, S., Kietzmann, T., Strom, S., Jungemann, K., Flieg, W., Michaopoulos, G., Serum-free, long-term cultures of human hepatocytes: Maintenance of cell morphology, transcription factors, and liver specific functions *Biochem. Biophys. Res. Commn.* **269**, 46, 2000.

94. Hino, H., Tateno, C., Sat, H., Yamasaki, C., Katayama, S., Kohashi, T., Aratani, A., Asahara, T., Dohi, K., and Yoshizato, K., A long-term culture of human hepatocytes which chow a high growth potential and express their difference phenotypes. *Biochem. Biophys. Res. Commn.* **256**, 184, 1999.

95. Thomson, J.A., Itkovitz Eldor, J., Shapiro, S.S., Waknitz, M.A., Sweirgiel, J.J., Marshal, V.S., and Jones, J.M., Embryonic stem cell lines derived from human blastocysts. *Science* **282**, 1145, 1998.

96. Shamblott, M.J., Axelman, J., Wang, S.P., Bugg, E.M., Littlefield, J.W., Donovan, P.J., Blumenthal, P.D., Huggins, G.R., and Gearhart, P.J. Derivation of Plurpopotent stem cells from cultured human primordial germ cells. *Proc. Natl. Acad. Sci.* USA **95**, 13726, 1998.

97. Hamazaki, T., Iiboshi, Y., Oka, M., Papst, P.J., Meacham, A.M., Zon, L.I., and Terada, N. Hepatic maturation in differentiating embryonic stem cell in vitro FEBS Lett. **487**, 15, 2001.

98. Thorgeirsson, S.S., Hepatic stem cells in liver regeneration. *FASEB J.* **10**, 1249, 1996.

99. Dunn, J., Tompkins, R., and Yarmush, M. Long term in vitro function of adult hepatocytes in a collagen sandwich configuration *Biotechnol. Prog.* **7**,237,1991.

100. Allen, J., Hassanien, T., and Bhatia, S. Advances in bioartificial liver devices, *Hepatology* **34**,447, 2001.

101. Landry, J., Bernier, D., Ouette, C., Goyette, R., and Marceau, N. Spheroidal aggregate culture of rat liver cells: Hystypic reorganization, biomatrix deposition and maintenance of functional activities. *J. Cell Biol.* **101**,914,1985.

102. Grompe, M., Overturf, K., Al-Dhalimy, M., and Finegold, M. Serial transplantation reveals stem cell line regenerative potential in parechymal mouse hepatoctes. *Hepatology* **24**, 256A 1996.

103. Rhim, J., Sandgren, E., Degen, J., Palmiter, R., and Brinster, R. Replacement of diseased mouse liver by hepatic cell transplantation. *Science* **263**, 1149, 1994.

104. Weglarz, T., Degen, J., and Sandgren, E. Hepatocyte transplantation into diseased mouse liver: Kinetics of parenchymal repopulation and identification of the proliferative capacity of tetraploid and ostaploid hepatocytes. *Am J. Pathol.* **157**, 1963, 2000.

105. Griffith, L., and Naughton, G. Tissue Engineering: Current Challenges and expanding opportunities. *Science* **295**, 1009, 2002.
106. Strain, A.J., and Neuberger, J.M., A Bioarticial Liver, State of the Art, *Science* Vol 295, 8 Feb. 2002.
107. Yarmush, M., Dunn, J., and Tompkins, R. Assessment of artificial liver support technology. *Cell Transplant.* **1**,323,1992.
108. Naruse, K., Sakai, Y., Nagashima, I., Jaing, G., Suzuki, M., and Muto, T. Development of a new bioartificial liver module filled with porcine hepatocytes cocultured in a microchannel flat plate bioreactor. *Inter. J. artif. Organs* **19**, 347, 1996.
109. Dixit, V., and Gitnick, G., The bioartificial liver: State of the Art. *Eur. Surg. Res.* **36**,71 1998.
110. Yang, S., Leonong, K. F., Du, Z., Chua, C.K., 2001, Review The Design of scaffolds for use in tissue engineering, Part 1, Traditional factors, *Tissue Eng.*, **7**, 6, 679-690.
111. Mooney, D., and Mikos, A. Growing new organs Sci. Am. April, 38, 1999.
112. Peters, S., and Mooney, D. Synthetic extracellular matrices for cell transplantation. *Mater Sci. Forum* **250**,43 1997·
113. Funatsu, K., Ijima, H, et al. Hybrid artificial liver using hepatocytes organoid culture, *Artificial Organ* 25(3): 194-200, 2001.
114. Hasirci, V., Berthjiaume, F., Bondre, S., Gresser, J., Trantolo, D., Toner, M., and Wise, D. Expression of liver-specific functions by rat hepatocytes seeded in a treated poly(lactic-co-glycolic) acid biodegradable foam, *Tissue Engineering* **7**, 4 2001.
115. Zeltinger, J., et al, Effect of Pore Size and Void Fraction on cellular adhesion, proliferation, and matrix deposition, *Tissue Engineering*, Vol. 7, Number 5 2001.
116. Mikos, A.G., et al. Prevascularization of porous biodegradable polymers, *Biotechnol. Bioeng.* 42, 716, 1993.
117. Wake, M.C., Patrick, C.W. and Mikos, A.G., Pore morphology effects on the fibrovascular tissue growth in porous polymer substrates, *Cell Transplant.* 3, 339,1994.
     Gion, T., Shimada, M., Shirada M., et al. Evaluation of a hybrid artificial liver using a polyurethane foam packed-bed culture system in dogs, *Journal of Surgical Research* 82, 132-136 1999.
118. Pierres, A., Benoliel, A., and Bongrand, P. Cell Fitting to Adhesive Surfaces: A prerequisite to firm attachment and subsequent events, *European Cells and Materials* Vol. 3, 2002, pp 31-45.
119. Gringell, D., Toberman, M., and Hackenbrock, I. Initial attachment of baby hampster kidney cells to an epoxy substratum. *J Cell Biol.* **70**: 707-713, 1976.
120. Shakesshieff, K., Cannizzaro, S., and Langer, R. Creating biomimetic micro-environments with synthetic polymer-peptide hybrid molecules, *Polymers for Tissue Engineering*, pp113-124, Moilley S Shoichet and Jeffery A Hubbell (Eds) VSP 1998.
121. Sussman, N.L., Gislason, G.T., Conlin, C.A., and Kelly, J.H. The hepatix extracorporeal liver assist device: initial clinical experience. *Artif. Organs* 1994;18:390–6.
122. Watanabe, F.D., Mullon, C.J.-P., Hewitt, W.R., Arkadopoulos, N., Kahaku, E., Eguchi, S., Khalili, T., Arnaout, W., Shackleton, C.R., Rozga, J., Solomon, B., and Demetriou, A.A. Clinical experience with a bioartificial liver in the treatment of severe liver failure: a phase I clinical trial. *Ann Surg* 1997;225:484–94.
123. Mazariegos, G., Kramer, D., Lopez, R., Shjakil, A., Rosenbloom, A., DeVera, M., Giraldo, M., Grogan, T., Zhu, Y., Fulmer, M., Amiot, B., Patzer, J. Safety observations in phase I clini cal evaluation of Excorp medical bioartificial liver support system after the first four patients. *ASAIO J.* **47**, 471, 2001.

124. Gerlach, J., Encke, J., Hole, O., Müller, C., Ryan, C., and Neuhaus, P. Bioreactor for a large scale hepatocyte in vitro perfusion, *Transplantation* **58**, 984, 1994.

125. Gion, T., Shimada, M., Shirada M., et al. Evaluation of a hybrid artificial liver using a polyurethane foam packed-bed culture system in dogs, *Journal of Surgical Research* 82, 132-136 1999.

126. Pierres, A., Benoliel, A., and Bongrand, P. Cell Fitting to Adhesive Surfaces: A prerequisite to firm attachment and subsequent events, *European Cells and Materials* Vol. 3, 2002, pp 31-45.

127. Kellogg, R.L., Nehring, R., Grube, A., Goss, D.W., and Plotkin, S., *Agricultural Productivity: Data, Methods, and Measures,* March 9-10, 2000, Washington DC.

128. From http://www.eps.gov/pesticides/primer.htm.

129. Personal communication, Lobster Institute, Orino, ME 2003.

130. Mülhmann, R. and Schrader, G., Hydrolyse der insektiziden Phosphorsäureester, *Z. Naturforsch.* 12b, 196 (1957).

131. Lawrence, L., Enzymes tackle another pesticide residue problem, *Australian Grain,* Nov. 2002.

132. LeJeune, K.E., Hetro, A.D., and Russell, A.J., Stabilizing nerve agent hydrolyzing enzymes, *Abstr. Pap. Am. Chem. Soc.,* (1997) 213 Meet., Pt.1, Envr239.

133. Havens, P.L. and Rase, H.F., Reusable immobilized enzyme/polyurethane sponge for removal and detoxification of localized organophosphate pesticide spills, *Ind. Eng. Chem. Res.,* (1993) 32, 10, 2254-58.

134. Hu, Z.C., Korus, R.A., and Stormo, K.E., Characterization of immobilized enzymes in polyurethane foams in a dynamic bed reactor, *Appl. Microbiol. Biotechnol.,* (1993) 39, 3, 289-95.

135. Huang, T.C. and Chen, D.H., Kinetic studies on urea hydrolysis by immobilized urease in a batch squeezer and flow reactor, *Biotechnol. Bioeng.,* (1992) 40, 10, 1203-09.

136. Storey, K.B., Duncan, J.A., and Chakrabarti, A., Immobilization of amyloglucosidase using two forms of polyurethane polymer, *Appl. Biochem. Biotechnol.,* (1990) 23, 3, 221-36.

137. Chakrabarti, A.C. and Storey, K., Enhanced glucose production from cellulose using coimmobilized cellulase and beta-glucosidase, *Appl. Biochem. Biotechnol.,* (1989) 22, 3, 263-78.

138. Bucke., C., Immobilized cells, *Phil. Trans.* R. Soc. B., 300, 369-389.

139. Lopez, S., Valero, F., and Sola, C., Immobilization of Cells Strategies in lipase production by immobilized Candida rugosa cells, *Appl. Biochem. Biotechnol.* 1996 vol. 59, no. 1, pp. 15-24.

140. Feijoo, G., Dosoretz, C., and Lema, J.M., Production of lignin peroxidase from Phanerochaete chrysosporium in a packed bed bioreactor with recycling, *Biotechnol.-Tech.* 1994 vol. 8, no. 5, pp. 363-368.

141. Ariff, A.B. and Webb, C., The influence of different fermenter configurations and modes of operation on glucoamylase production by Aspergillus awamori, Asia-Pacific *J. Mol. Biol. Biotechnol.* 1996 vol. 4, no. 3, pp. 183-195.

142. Targonski, Z. and Pielecki, J., Continuous semi-solid cultivation for the production of cellulase by *Trichoderma reesei* mutants using a polyurethane foam carrier and a liquid medium, *Acta-Biotechnol.* 1995 vol. 15, no. 3, pp. 289-296.

143. Sun, Y., Li, Y.L., Yang, H., Bai, S., and Hu, Z.D., Characteristics of immobilized Rhizopus oryzae in polyurethane foam cubes, *Biotechnol.-Tech.* 1996 vol. 10, no. 11, pp. 809-814.

144. Hall, D.O. and Rao, K.K., Immobilized microalgal systems., *Br. Phycol. J.* 1990. vol. 25, no. 1, p. 89.

145. Domingo, J.W.S., Radway, J.C., Wilde, E.W., Hermann, P., and Hazen, T.C., Immobilization of Burkholderia cepacia in polyurethane-based foams: Embedding efficiency and effect on bacterial activity, *J. Ind. Microbiol.-Biotechnol.* 1997 vol. 18, no. 6, pp. 389-395.

146. Tuerker, M. and Mavituna, F., Production of cellulase by freely suspended and immobilized cells of Trichoderma reesei, *Enzyme-Microb.-Technol.* 1987. vol. 9, no. 12, pp. 739-743.

147. Thomson, Hydrophilic Polyurethane for the Delivery of Skin Care Ingredients, US patent Application No. US 2002/0182245 A1, 2002.

148. Takano, H. et al, (The Artificial Heart: Present Status and Future Prospects), *Nippon Geka Gakkai Zasshi,* 92(9) Sept 1991 1258-62.

149. Flory, P.J., *Principles of Polymer Chemistry,* Cornell University Press, Ithica, NY 1953.

150. Kuhn, W., Hargitay, B., Katchalsky, A., and Eisenberg, H., *Nature,* Vol. 165, April, 1950, pp 514-516.

151. Tanaka, T., Nishio, I., Sun, S.T., and Ueno-Nishio, S., *Science* Vol. 218, Oct. 1982 pp 467-469.

152. Tanaka, T. US Patent No. 5,100,933, Collapsible gel compositions, 1992.

153. Baughman, R. H., Cui, C., et al., *Science,* Vol 284 may 1999, pp1340-1344.

154. Thomson, T., US Patent No. 6,117, 296, *Electrically Controlled Contractile Polymer,* 2000.

Index

A

Absorption
  in extraction process, 5, 67
  of fluids, 2–3
  reservoir capacity and, 57–58
Acids, filtration removal during extraction, 69
Acrylics, in artificial muscle development,
    162–163
Activated charcoal
  absorptivity, 5, 62, 67
    polyurethanes vs., 71–72
  in aquaculture, 99
Activation energies, 54
Active ingredients
  within a matrix, 85
  in skin care products, 158–159
Additives, certification standards, 46–47
Adhesion
  in biofiltration systems, 108
  cellular (*see* Cell adhesion/attachment)
Adsorption, 5
  of albumin, 114–115
  in biofitration systems, 108–109
  in extraction process, 67
  as immobilization, 150–152, 157
  of proteins, 59, 113–116
    avoiding in implantable medical devices,
      117–119
Aerobic process, in biochemical conversion,
    88–91
AGBs, *see* Attached growth bioreactors (AGBs)
Aging process, of skin, 158
Agriculture
  pollutants from, 4, 61, 87, 96, 105
    pesticides, 148–149
  polyurethane applications, 160–166
    artificial muscle development, 161–164
    high-value plants, 160–161
Airborne pesticides, 149
Air flow
  in biofiltration, 97
  as flow-through system test, 29, 43–47, 97
Air pollution
  biofiltration of, 11–12
    hydrophilic-hydrophobic composites,
      101–103

    reticulated polyurethanes, 97–98,
      101–103
  extraction processes, 5–6, 70
  microorganism treatment of, 97–98
  trends, 62
Alachlor, 149
Albumin, adsorption of, 114–115
Alcohol-grafted polyurethane, 71
Alcohol groups, in polyols, 8, 21
Aldrin, 149
Algae
  in biofiltration systems, 108–109
  immobilization of, 157
  recycling role, 4, 9
Aliphatic isocyanates, 50, 69, 71, 150
American Society for Testing and Materials
    (ASTM), 38, 43, 45
Amines
  biodegradability and, 25, 47, 122
  in enzyme immobilization, 151
Amino acids, biodegradability and, 122
Ammonia
  in aquaculture, 99–100
  biochemical conversion, 90–91
  as water pollutant, 86, 88
Amyloglucosidase, 60
Anaerobic process, in biochemical conversion,
    88–90, 104–105
Anchorage-dependent cells
  hepatic, 133, 137
  viability of, 124–125
Antibead, 3
Antibody-antigen coupling, in medical assist
    devices, 138
Apoptosis, 124
Aquaculture, hydrophilic polyurethanes for,
    99–101
Aquifers, for extraction, 66–67
Archaea organisms, water recycling role, 90
Aromatic hydrocarbons
  extraction from air, 56
  as pollutant, 62, 69, 71, 150
Artificial liver, 14–15, 129–131
  current clinical activity, 140–146
  extracorporeal, 129–131, 134, 140–141
  scaffold-based, 129, 140–146

F

O

P